DYER

INTENSITY COILS

How Made and How Used

Elibron Classics
www.elibron.com

Elibron Classics series.

ISBN 1-4212-6335-1 (paperback)

This Elibron Classics Replica Edition is an unabridged facsimile of the edition published in 1891 by Perken, Son & Rayment, London.

PRESS OPINIONS.—"*Intensity Coils: How Made and How Used.*"

ENGLISH MECHANIC.

"In the preface it warns us that it is not put forth as a scientific exposition of the matter; yet, for all that, many of the explanations are clear and good, and directions for experiments easily to be followed."

ELECTRICAL ENGINEER.

"'Intensity Coils: How Made and How Used.' By Dyer. Sixteenth Edition. London: Perken, Son and Rayment. A book that has reached a sixteenth edition, and which has been before the world for many years, must contain information that is wanted. The Ruhmkorff coil has become one of the most interesting pieces of apparatus in electrical engineering in its comparatively recent development, known as the transformer. This book shows very clearly the historical views held about the coil, and from the amateur constructor's point of view is most valuable, in that it explains clearly how to make and how to use a coil. A great many lecture experiments are described."

CHEMIST AND DRUGGIST.

"'Intensity Coils.' By 'Dyer.' London: Perken, Son and Rayment, 99, Hatton Garden. 1s. This is one of those valuable little popular manuals which convey a sound elementary knowledge of an important subject in a concise manner. It describes the principal batteries, deals fully with making coils, and illustrates their use by numerous experiments. There are in addition brief explanations of the electric light, the telegraph, telephone, phonograph, &c."

ARCHITECT

"'Intensity Coils.' This is the title of one of the publications of Messrs. Perken, Son and Rayment, of Hatton Garden. It is a "beginner's guide to electricity," describing the way to make batteries, bells, coils, electric light, telegraphs, phonographs, telephones, &c. It is already passing through its sixteenth edition, or the 128th thousand. One of the merits of this brochure is that it does not require an expert to understand it.'

PHOTOGRAPHIC NEWS.

"Messrs. Perken, Son and Rayment, the sixteenth edition of 'Intensity Coils, being a beginner's guide to electricity, describing the way to make batteries, bells coils, electric light, telegraphs, phonographs, and so on."

PHOTOGRAPHY.

"'Intensity Coils.' Perken, Son and Rayment, 99, Hatton Garden, E.C. This forms one of the firm's well-known publications, and is a sort of beginner's guide to electricity, describing the way to make batteries, coils, lights, bells, telegraphs telephones, phonographs, &c."

CHEMICAL NEWS.

"The induction coil has now come into such general favour, and is used for such varied purposes as a medical instrument, a means of scientific research, and an amusing toy for winter's evenings, that several treatises have lately appeared, describing the best methods of its manufacture and management. The object of the little work before us is explained in the following manner in the preface:—

"What is an intensity coil? How does it differ from other coils? How is it made? What will it do? These and similar questions are being continually asked, and to furnish intelligible replies to them the following pages have been written. It is not a scientific treatise that is here offered to the public, but simply the necessary instructions that those who want to make or use intensity coils desire to obtain.'

CITY PRESS.

"'Intensity Coils.' By 'Dyer' (Perken, Son and Rayment). When a handbook, as for instance the one now demanding attention, retains its position for nearly a quarter of a century, few will be inclined to dispute the contention that it is a work possessing considerably more backbone than the generality of such publications. The present book was first published in the year 1867, and it is now in its 16th edition. In its pages will be found, given in an eminently readable form, much that will serve to instruct the young idea concerning the electric light, the telephone, the microphone, and other wonders which, more or less, are associated with the great American inventor, Edison. With all the confidence we displayed when twenty years ago we noticed the first edition, we can heartily commend this handbook to the notice of our readers as a publication 'which will be of service to all persons' engaged in the interesting experiments about which 'Dyer' has so much to say."

BRITISH JOURNAL OF PHOTOGRAPHY.

"Messrs. Perken, Son and Rayment have issued a small book—'Intensity Coils; How Made and How Used.' It forms an excellent beginner's guide to electricity, describing the way in which batteries, coils, &c., are made. The previous editions of this work have evidently found much favour, for the present is the sixteenth edition. It contains numerous illustrations."

PRESS OPINIONS.—*"Intensity Coils: How Made and How Used."*

SPORTING AND DRAMATIC.

"'Intensity Coils.' Messrs. Perken, Son and Rayment publish the sixteenth edition of a capital little handbook on this subject by 'Dyer.' Electric lighting, electric bells, the telegraph, the telephone, the microphone, the phonograph, &c., are pithily and comprehensibly explained. The work, which has been revised to date, is published at one shilling, so that a practical knowledge of applied electricity up to date is within the reach of everybody."

OVERLAND MAIL.

"Messrs. Perken, Son and Rayment's useful guide to electricity for Beginners, entitled 'Intensity Coils: How made and How Used,' is now passing through its sixteenth edition (128th thousand). It describes the construction and method of using galvanic batteries, while there are chapters on electric lighting, the telegraph, telephone, and Phonograph. The chapters are subdivided into paragraphs, which are headed in large type. To the amateur electrician this work is invaluable."

PICTORIAL WORLD.

"'Intensity Coils: How Made and How Used.' This is the third of a series of scientific primers brought out by the well-known firm of Perken, Son and Rayment, 99, Hatton Garden, E.C. It should prove of great use to all those interested in electricity; it is carefully complied and brought up to date, containing much useful information on electric lighting, bells, motors, the telephone, microphone, and phonograph. Directions are fully given for seventy or eighty instructive and amusing experiments; and we predict a large sale for this, the latest edition."

BUILDERS' REPORTER AND ENGINEERING TIMES.

"'Intensity Coils.' The scientific information is of an advanced character; but the reader is introduced to it by a preliminary clear and simple explanation of all the rudimentary steps, technical terms, &c. The application of electricity to the various well-known purposes is explained, and the latest projects, and, among these, 'seeing by electricity' are noticed.

MERCANTILE GUARDIAN

"'Intensity Coils.' Sixteenth edition. By 'Dyer.' London: Perken, Son and Rayment. 1s. Electricity, like photography, is no longer the exclusive pursuit of the professional expert; for the ubiquitous amateur has made it his own, as he has most other things, thus necessitating elaborate handbooks for his particular benefit. The induction coil is much in favour, as it can be applied in so many ways—as a medical instrument, for scientific research, and as part of a drawing-room entertainment. We can advisedly recommend this most useful little handbook, as it contains in a condensed form just the information required. Other branches of the science are treated upon, such as electric lighting, bells and telegraph; the telephone, microphone and phonograph."

INVENTION.

"'Intensity Coils: How Made and How Used,' Perken, Son and Rayment, 99, Hatton Garden. This is a very useful beginner's guide to electricity, although its title hardly gives a fair idea of its contents. It deals with electric lighting, bells, telegraphs, motors, the telephone, and the microphone, while the phonograph is also noticed, on the ground that its invention was an outcome of the investigation of the phenomena connected with the telephone. This being the 16th edition (128th thousand), the fact that the work supplies a public want may be safely assumed."

CIVILIAN.

"'Intensity Coils: How Made and How Used;' also electric light, telephone, phonograph, &c. By 'Dyer.' Messrs. Perken, Son and Rayment, 99, Hatton Garden. This is a little handbook which will be of great use to amateurs engaged in their interesting experiments. It is concisely written, the explanations are clear and good, and it is evidently the work of one who has himself done the work. Students will have to find out what is an intensity coil? How does it differ from other coils? How is it made? What will it do? Besides a mass of other information, including seventy or eighty amusing and instructive experiments. All who have a scientific bent will welcome this little book.

HOME NEWS.

"'Intensity Coils: How Made and How Used.' By 'Dyer,' (Perken, Son and Rayment. 1s.) Describes the manufacture of batteries, bells, coils, electric light, telegraphs, telephones, phonographs, &c., in a crisp, clear, understandable style, which the amateur will value very highly. It is not a book of 'pretension,' but no better testimonial as to its merits as a popular handbook to the fascinating subject of which it treats could be mentioned than the fact that the little volume is now passing through its sixteenth edition, or 128th thousand,

INTENSITY COILS:

How Made and How Used,

BY

"DYER."

WITH A

DESCRIPTION OF THE ELECTRIC LIGHT—
ELECTRIC BELLS—ELECTRIC MOTORS—THE TELEPHONE—
THE MICROPHONE—AND THE PHONOGRAPH.

SEVENTEENTH EDITION.

London:

PUBLISHED BY THE PROPRIETORS,

PERKEN, SON & RAYMENT, 99, HATTON GARDEN,

AND SOLD BY ALL BOOKSELLERS AND OPTICIANS.

1891.

WOODFORD FAWCETT & Co.
PRINTERS, DORSET WORKS, SALISBURY SQUARE, E.C.

Preface to the Seventeenth Edition.

A NEW EDITION of this work being required, the opportunity has been taken to make some additions to it, which, it is hoped, will render it even more worthy of the place it has attained. In addition to describing the construction and method of using Galvanic Batteries and Intensity Coils, other branches of the science receive attention—viz., Electric Lighting—Electric Bells—Electric Telegraph—Electric Motors—the Telephone—the Microphone—the Phonograph.

"DYER."

June, 1891.

INTENSITY COILS:

How Made and How Used.

BY "DYER."

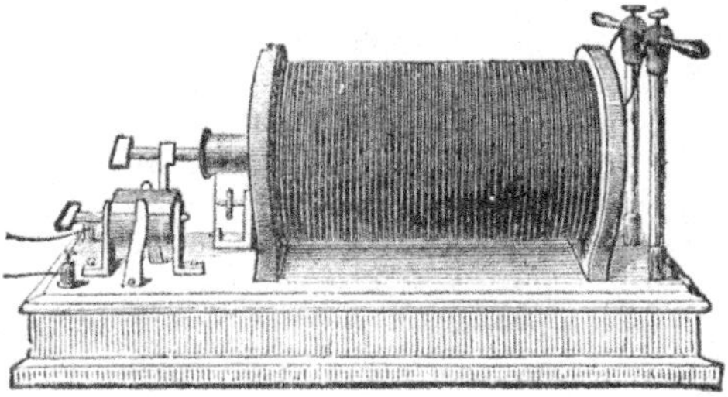

PRELIMINARY EXPERIMENTS.

1. Discovery of Current Electricity.—When the youthful pupil of Galvani accidentally brought the legs of a dead frog into an electric circuit, he witnessed the manifestation of an unknown force. Neither he who first saw the phenomena, nor they who afterwards repeated the experiments, had the remotest idea that the things they thought so curious were, in fact, only the germs of extraordinary discoveries. If these first manipulators in Current Electricity could not look *forward* to the stupendous result now achieved by its aid, it is almost equally difficult for us to look back upon their origin. We, whose statues and ornaments are produced by electric agency, whose bells are rung by its aid, who see rocks blasted and the obstructions in ports and rivers removed by its instrumentality, and whose thoughts can be conveyed, as it were, instantaneously, thousands of miles over rocky chains, desert wastes, and all but fathomless oceans by the same wondrous power, even we can hardly realise that all this is to be traced to the accident—perhaps mischievous freak—of a doctor's boy.

Without recurring to the physiological effects that once comprised all that was known of Galvanism, it will be necessary to advert to a few primary principles in order that the complex phenomena specially to be treated of may be properly understood.

2. Definition of Galvanism.—Galvanism may be defined as a manifestation of Electricity in a current form, produced by a chemical action.

Galvanism is an invisible agent; it is not to be apprehended by the touch; any or all of our senses may be brought to bear upon it, and yet it eludes our observation; it is only known to us by its effects. But it is a real force; it is one in nature with that mighty power which, in the form of lightning, rends the rocks, uproots the firmest trees, and has ever been regarded by man as a fitting emblem of the Deity.

3. Manifestation of Galvanism.—Prepare a glass vessel, and partly fill the same with water made strongly sour by the addition of a little sulphuric acid. Into this plunge two plates of metal, attached together by a copper wire, one metal plate being amalgamated zinc, and the other platinised silver. *Fig.* 3. Noticeable results follow. The surface of the platinised silver becomes coated with minute bubbles, which increase in size, and are eventually disengaged, and rising through the liquid, escape into the air. The whole of the liquid in the neighbourhood of the platinised silver plate soon assumes a state of effervescence, while that surrounding the zinc is free from disturbance. Distinct and marked though these effects are, yet they represent but a portion of the changes that are taking place. The zinc plate, by its superior affinity for oxygen (one of the elements of the water in the glass jar), decomposes the film of water immediately in contact with it, the result being the formation of an oxide of zinc over its surface. The other element of the water, hydrogen, eventually passes over to the platinised silver plate, and appears there in the form of the bubbles described. The oxide of zinc formed on the surface of the zinc plate is dissolved off by the action of the sulphuric acid in the water, forming sulphate of zinc, a fresh surface of zinc is exposed, another film of water is deprived of its oxygen, and another coating of oxide of zinc formed, and another equivalent of hydrogen is given up at the surface of the platinised silver plate. This will go on as long as there is any zinc left, or any capability in the water to dissolve the sulphate of zinc so formed.

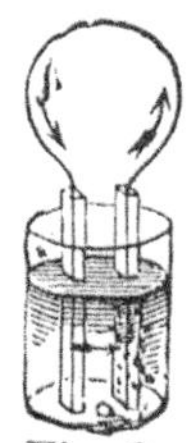

Fig. 3.

But the results do not stop here; not only has a new compound been formed at the expense of the zinc and the water; not only has one of the elements of the water been given up into the atmosphere in the form of a gas, but an entirely distinct although invisible force has been evolved. A current of Galvanic electricity has been set up, starting from the zinc plate, passing through the fluid to the platinised silver plate. The electricity does not rest here; it proceeds up the silver plate to the attaching wire, and then through this wire back again to the zinc plate.

4. Evidences of the existence of Galvanism.—But what proof, it may be asked, can be given that there is this invisible intangible current flowing through the wire? How is its existence to be known? Another experiment will afford the evidence required. Prepare a glass vessel, partly filling the same with acidulated water (**3**)*, and before plunging in the zinc and silver plates, separate the wire that connects the plates together, as shown in *Fig.* 4. The effervescence before seen (**3**) as surrounding the silver plate, will then be looked for in vain—all will be quiescent.

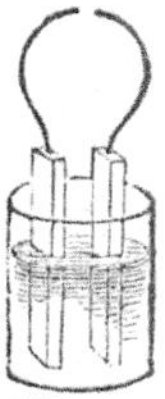

Fig. 4

If now the ends of the two wires be brought into contact, the action observed in the former experiment (**3**) will immediately commence. Clearly, therefore, the effervescence seen in the fluid does not represent all that is taking place; for it is evident that in some way or other a *connection* between the two plates is essential to the existence of any action at all.

Further, let the wires be again separated, and all the action will cease. Bring the wires together again, but interpose between their ends a film of paper, ivory, gutta-percha, or ebonite. By such an arrangement the plates are still united, but no action results. Separate the wires again, removing the intervening material, and replace it by a piece of brass, silver, or any other metal, and the action immediately recommences as at first. Thus it is seen that not only must the plates be joined together, but they must be united by a material analogous to a metal, and not by one like paper, ivory, gutta-percha, &c. It is evident something must flow through the connecting wire, and that unless the wire through its entire length be of a metallic kind, it is as though it were severed or broken wire. It is also obvious that some substances possess the property of being conductors of what passes through the wire, and other substances do not possess that property.

5. Influence of the Connecting Wire on a Magnet.—The foregoing will be carried a step farther if it be found that the wire connecting the two plates possesses properties, while the action is going on, that do not belong to it under other circumstances.

Prepare the apparatus as used in the preceding experiment (**4**); first elongating the connecting wire, as shown at *Fig.* 5. The

* These figures within brackets in the body of the text refer to the paragraphs bearing similar numbers.

plates are to be put into the glass vessel before the acidulated water is poured in, and the connecting wire must be arranged so that it may come over an ordinary mariner's compass needle, *Fig.* 5. This wire must coincide in direction with the needle, pointing north and south, and be about half an inch above it. No result whatever will follow; the wire, although attached to the plates, will have no influence on the compass needle any more than it would have had if it were detached from the plates.

Let the acidulated water be now poured into the glass vessel, and the wire will be found to have acquired a marvellous power; the compass needle will be forced out of its position and stand oblique, as shown by the dotted lines, *Fig.* 5. This is not a transient movement; the needle will permanently remain in a position more or less oblique to the direction of the wire. If now the connecting wire be cut through so as to separate it (**4**), all its powers will disappear, the compass needle will return to its original position, and all action will cease in the glass vessel.

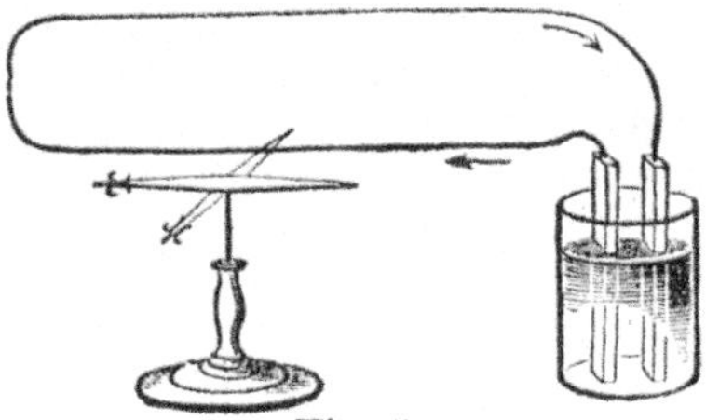

Fig. 5.

This experiment clearly proves that, associated with the action going on in the glass vessel, there must be some power or influence traversing the connecting wire, and communicating to it properties that it did not previously possess. It also shows that such properties only exist whilst the action in the glass vessel is going on. This power or influence is called Galvanic or Current Electricity.

The connecting wire, under these circumstances, possesses other properties than those just described, but it will not be necessary to refer to them for the purposes of this paper.

6. Œrsted's Discovery.—The experiments just detailed appear to be very simple, but they have sufficed to open a wide field for scientific research. For a long time it had been doubted, and even denied, that an electrified wire (as a wire conveying an electric current is called) and a magnetised needle had any mutual relationship. It was not until the year 1819 that it was ascertained that such relationship did exist, and it is to Professor Œrsted, of Copenhagen, that the world is indebted for the discovery.

7. Galvanometers.—One of the early results of Œrsted's great discovery was the construction of instruments called Galvanometers (or more properly Galvanoscopes), for detecting the presence of electrical currents. A Galvanometer consists essentially of a magnetised needle suspended by a fibre or balanced on a point with one or more strands of wire passing round it, the two ends of such wire being represented by two brass caps or binding screws, *Fig*. 7.

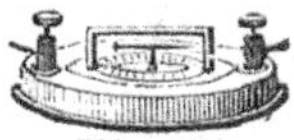

Fig. 7.

When the arrangement is to be subject to examination, it has to be attached to the binding screws of the Galvanometer, and if there be any electric current traversing it, the needle will be acted on and deflected from its position (**5**). In arranging a Galvanometer for use, it must always be so placed that the needle, when at rest, may coincide in *direction* with the wire that surround it.

8. Action of Electrified Wires on other Wires.—An electrified wire not only possesses the power of forcing a magnetic needle out of its natural position, but it can affect also a contiguous wire, which phenomena are well seen in the following experiments.

Place a plate of zinc and one of silver in an empty glass vessel (**5**). Fix a wire into the binding screws of the Galvanometer (**7**), and arrange the Galvanometer so that this wire shall be parallel to the wire connecting the two plates together, as shown in *Fig*. 8. These two wires must not touch each other, and should be a ¼in. apart. Two circuits are thus formed, one through the plates in the glass vessel and the wire that connects them together, and the other through the Galvanometer and the wire uniting the two binding screws. Let the acidulated water be now poured into the glass vessel, and the needle of the Galvanometer will be instantly forced out of its position, or, as it is usually termed, deflected (**5**). Thus we see that one electrified wire can originate a similar current in a wire near to, but not in any way connected with it. The electrical current thus set up is called an induced or Secondary Current, whereas the current in the wire connected with the plates in the glass vessel is called the Battery or Primary Currrent.

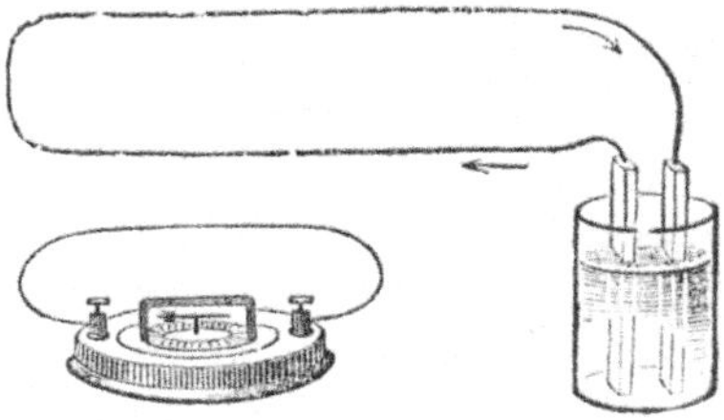

Fig. 8.

9. The Simple Battery.—The plates and glass jar and aciduated water used in the preceding experiments are designated a Simple Battery, by which name they will now be called.

10. Influence of an Electrified Wire on an Iron Bar.—Prepare the Simple Battery (**9**), and instead of using a straight wire to connect the plate together, twist it into the form of a helix, *Fig.* 10, and for this experiment the wire should be *insulated*, that is, covered with cotton, silk, or gutta-percha. Place within this helix a small bar of iron, and then pour the acidulated water into the glass vessel. If the iron bar be now tested by applying iron filings or small tacks to its projecting ends, it will be found to have become magnetic. Remove the bar from the helix, and it will at once lose all its magnetic property. Replace it in the helix, and it again becomes magnetic; take the plates and helix and iron bar out of the solution in the glass vessel, and again all the magnetism of the bar disappears.

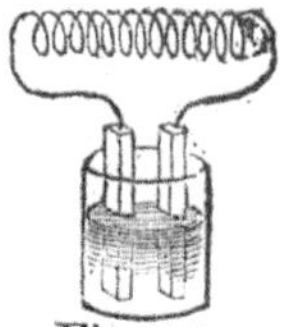

Fig. 10.

Thus it is evident that if a wire carrying a current of electricity be wound round an iron bar, temporary magnetism is communicated to the bar, which magnetism exists only as long as the iron bar is surrounded by the coil of wire and the electric current continues to flow.

11. Electro-Magnets.—The magnetic state is called Induced Magnetism, to distinguish it from the permanent magnetic condition of *steel* bars, which we are familiar with in the form of horse-shoe magnets. If instead of using a feeble battery, such as has been described, one of a more powerful description be employed, and if the iron bar be bent, as shown in *Fig.* 11, and several convolutions of wire be wound on it, an induced magnetic force can be obtained sufficient to sustain a weight of hundreds of pounds. Such a bar of iron, whether straight or bent, is termed Electro-magnet.

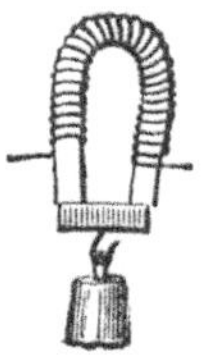

Fig. 11.

12. Action of a steel Magnet on a Helix of Wire.—The connection which exists between the phenomena of Magnetism and that of Electricity is of the closest description. We have seen the influence of an electrical wire on an iron bar (**10**), the Electricity in the one including Magnetism in the other, and the converse of this experiment is equally striking.

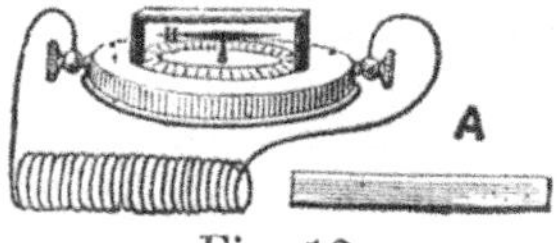

Fig. 12.

Attach a wire helix (**10**) to the Galvanometer, as shown in *Fig.* 12. No effect will be produced on the Galvanometer by the helix; but if a permanent steel magnet A be passed into it, the Galvanometer will be at once influenced, and the needle deflected. Thus it is seen that an electrical current can be excited in a wire by the action of a permanent magnet, as it was seen (**10**) that magnetism can be excited in an iron bar by the influence of an electrified wire.

12a. Luminous Effect of Magnetically Induced Electricity.—This induced current may be made to manifest itself not only by deflecting the needle of the Galvanometer—but also by a bright spark in the following way:—Take a short rod of soft iron, and wind round it some insulated copper wire—one end of the wire terminating in a copper disc, about the size of a penny—and the other end bent over so as to rest lightly on the disc. Now, taking hold of one end of the rod, bring the other end down on to the pole of a good size horse-shoe magnet—the concussion will cause a separation to take place between the copper disc and the wire which rests on it, simultaneously with the excitation of a current of electricity in the wire by the inductive action of the magnet—and a bright spark will be seen. Another luminous discharge will take place on removing the iron rod from the magnet.

13. Increased Effects resulting from Increased Current Action.—All the effects produced by the Simple Battery described in the preceding experiments may be increased by employing a more powerful battery, but the difference will be in *extent* only, the phenomena themselves remain the same.

Although in performing these experiments it does not at first sight seem that much has been done, yet it is from the principles thus demonstrated that the wonderful science of Telegraphy has proceeded. However complicated and surprising the present condition of that part of science is, Telegraphy is but the application of the facts shown in the experiments just detailed to some of the requirements of daily life.

14. Current, or Dynamic Electricity.—The electricity evolved by a Galvanic battery is designated Current or Dynamic Electricity (**5**), and requires for its conveyance a complete circuit of some conducting substance or substances from the plate emitting the current back to the same plate again.

15. Static Electricity.—There is another condition of Electricity called Static Electricity that should be adverted to, as both states are manifested by the Intensity Coil. The Static condition of Electricity differs from the Current in that it may be regarded as a state of quiescent accumulation. It is associated with some *surface* which is then said to be *charged*, and such surface will emit sparks whenever any conductor is presented to it. A familiar experiment will illustrate this. Take a stick of sealing wax, and rub one end briskly with a piece of fur or flannel, *Fig*, 15. When this has been done, the wax will become electrically excited, and if the knuckle be passed over the surface,

Fig. 15.

a series of small sparks will be seen, accompanied by a cracking noise. This is Electricity in the Static condition.

GALVANIC BATTERIES.

Fig. 20.

16. Galvanic Batteries.—It will be proper now to give some explanation of the various kinds of Galvanic Batteries that are generally employed. This will prevent any further references to specific forms of Battery when the experiments with the Intensity Coil are described. The term "the Battery" will then be used, and that phrase will be taken to express the particular kind, as well as the number of cells that may have been selected.

17. Elements of a Galvanic Battery.—To constitute a Galvanic Battery it is necessary (**3**) to have two dissimilar substances or surfaces, both being conductors of electricity, and they must be placed in a fluid that will act more energetically on one than on the other. The substance of surface *less* acted upon is called the *Negative element*, and that *more* acted upon the *Positive element* of the battery. The substances or surfaces may be flat or circular; they may be near together or more widely asunder; they may be both of them metallic, or only one of them a metal;—such variations only affect the *force* of the electric current evolved. The materials used as the elements of Galvanic Batteries are usually such as these: Copper and Zinc, Silver and Zinc, Platinum and Zinc, Carbon and Zinc, Cast Iron and Zinc; and for the exciting fluids, Sulphuric and Nitric Acids, Bichromate of Potassa, and Sulphate of Copper. Many other combinations can be employed for evolving electric currents, but results so obtained are more interesting in a scientific than in a practical point of view. Therefore the Cruikshank's, Daniel's, Smee's, Grove's, Callan's, Bunsen's, and the Bichromate batteries, which are those that require the materials mentioned, will alone be described.

One only of the elements of the battery is acted upon by the exciting fluid, and dissolved (or worn away, as the usual phrase is). This is the Positive *element*, while the other, which is not acted upon or worn away, is the Negative *element*.

18. Poles of a Galvanic Battery.—Much confusion exists in the minds of many persons with reference to the terms the Positive and Negative *elements* of a battery and the Positive and Negative *poles* of a battery. This is readily dissipated by observing that the term ELEMENT is applied to that part of the plate of a battery that is *within* the exciting fluid, such being the active part of the plate; and the term POLE to that part of the plate that is *without* the fluid, and which becomes the conductor only. The term POSITIVE is intended to signify *that* from whence the current proceeds, while the NEGATIVE signifies *that* which the current enters. By a reference to the accompanying illustration, it will be seen that the same plate of a battery is the *Positive* element and the *Negative* pole. By following the course of the arrows, *Fig.* 18, it will be found that the current flows from one plate through the fluid and up the other plate. It then passes from the end of that plate through the wire that connects the two together to the top of the plate from whence it originally started. Thus it is that *each* plate of a battery has opposite terms applied to it, one part being called Positive and the other part Negative.

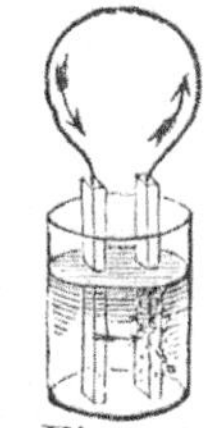

Fig. 18.

19. Electrodes.—Dr. Faraday disuses the word *pole* altogether, and calls the ends of the elements that are outside the exciting fluid ELECTRODES, the paths or ways of the electricity; and to express the *emitting* or Positive Electrode, he prefixes the syllable AN, and to express the *receiving* or Negative Electrode, the syllable CATH. So that in the language of that distinguised man, the upper end of the Negative element of the battery (usually called the Positive *pole*) is the emitting or An-electrode, and the upper end of the Positive element (usually called the Negative *pole*) is the receiving or Cath-electrode. For convenience of pronunciation these two words are shortened to Anode and Cathode: and in deference to so distinguished an authority these terms will be used throughout this book, instead of the words Positive pole and Negative pole.

20. Compound Batteries.—In the foregoing remarks, a simple or single cell battery only has been described. But the principles detailed belong equally to any multiple of the single cell, forming what is then called a compound battery.—*Fig.* 20.

21. Quantity Arrangement of a Battery. — In uniting single cell batteries to form a compound battery, the anode of each single cell battery may be attached to one metallic bar or wire, and the cathode of each single cell to another similar bar or wire. This is called the quantity method of arrangement, and the effects obtained from such a compound battery will be similar to those that would be produced by a single cell battery, the plates of which were equal in surface to the sum of all the plates in the compound battery. Such a method of connection converts, in fact, a *number* of small batteries into *one* large one.

22. Intensity Arrangement of Battery.—If it be desired to give to the current of electricity evolved from any single cell battery a greater power of overcoming obstacles, then other batteries must be united with it in such a manner that the current from the one battery will circulate *through* all the batteries that are added to it. This is called the intensity arrangement, and it requires that the anode of one battery be united to the cathode of the next, and so on through the entire series employed.

23. Necessity for Uniformity of Size in Batteries.—In uniting single batteries together to form an intensity arrangement, care should be taken to use batteries whose negative elements are of *equal* size. If this be not attended to, the power of the series will be reduced to what it would be were all the batteries the size of the *smallest* one. The absolute power of any galvanic battery, all other things the same, depends upon the area of the surface of the *negative* element.

24. Electrodes of Compound Batteries.—When a compound battery is arranged as a quantity battery (**21**), any part of the bar or wire to which the anodes of the single batteries are united will form the anode of the compound battery; and any part of the bar or wire to which the cathodes of the single batteries are attached will be its cathode. When single batteries are arranged for an Intensity Battery (**22**), the *unconnected* cathode of the first battery will be the cathode of the series, and the *unconnected* anode of the last battery will be the anode of the series. To attach a Galvanic Battery, whether a simple or a compound form, to any piece of apparatus, it is only necessary to carry wires from the two electrodes to the apparatus that is to be employed.

25. Continuity of Circuit.—The effects producible by a Galvanic Battery are manifested in the passage of the current from one electrode to the other (**4**): and, therefore, in arranging a Galvanic Battery and apparatus, an uninterrupted conductor of some kind or other must be provided between the two electrodes.

BATTERIES, HOW CONSTRUCTED.

26. Cruikshank's Battery.—This Battery is only used as a compound Battery (**20**). It consists of plates of zinc and copper united together in pairs, and fitted into groves in the sides of a wooden trough. A suitable space is left between the pairs of plates, and this space forms the cell for receiving the exciting fluid. Cruikshank's Battery is excited with a diluted solution of sulphuric acid, with a slight addition of nitric acid, but it is not a very convenient form for experimental purposes. It is used for telegraphic purposes, and also for medical galvanism.

27. Daniel's Battery.—This admirable Battery consists of a copper cylinder, in which is placed a cylinder of porous earthenware, and in this a zinc rod, *Fig.* 27, the zinc being the *positive* and the copper the *negative* element. Two exciting solutions are required for this battery: one a saturated solution of sulphate of copper, and the other a solution of sulphuric acid (commercial oil of vitriol), in the proportion of seven parts of water, by measure, to one part of sulphuric acid. The solution of sulphate of copper is placed in the copper cylinder, and the diluted sulphuric acid in the porous cylinder. The porous cylinder keeps the two solutions separate, and at the same time allows the electric current to pass through its sides. A binding screw attached to the zinc rod, and one attached to the copper cylinder, form the cathode and anode (**19**). It is not necessary to have any containing vessel for this battery; the copper cylinder being closed at the bottom performs that office. This battery is very effective in use, remarkably constant, and gives a current of considerable intensity.

Fig. 27.

28. Smee's Battery.—This form of battery has perhaps been more extensively used than any other galvanic arrangement. It consists of a plate of platinised silver for the *negative* element, with two plates of zinc for the *positive* element, *Fig*. 28. The platinised silver plate is usually attached to a wooden bar; and the zinc plates, being placed one on each side of the same, are kept in the required position by a metal cramp passing over the top of the bar.

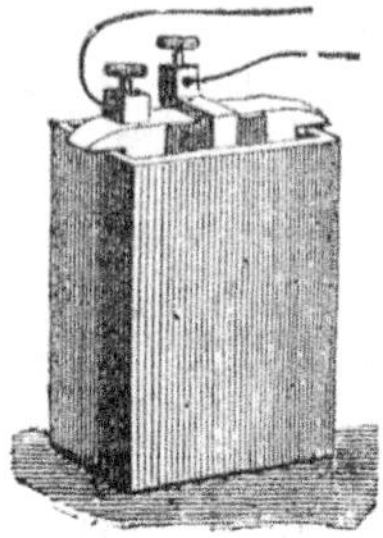

Fig. 28.

A binding screw, passed through the wooden bar, and attached to the silver plate, forms the anode, and a similar binding screw, on the cramp that holds the zincs to the bar, is the cathode. An earthenware or other suitable containing vessel is required for this battery, and it is excited by a solution of sulphuric acid (commercial oil of vitriol) in the proportion of 7 parts of water by measure to 1 part of sulphuric acid. This battery is admirably adapted for electro-depositions, and also for general galvanic experiments; but it is not the most suitable kind for producing electric light, nor for the intensity coil. It is easily managed, is tolerably constant, and requires only one exciting fluid; and, therefore, porous cells are dispensed with.

29. Grove's Battery.—The elements of this battery are platinum for the *negative* and zinc for the *positive*. This battery also requires a containing vessel to hold the entire arrangement, and an inner cell to hold the platinum plate only. This inner cell, like that in Daniel's battery (**27**), must be of porous earthenware, which will admit the passage of the galvanic current through its sides, but will not allow the exciting fluids to mingle. When the zinc plate is placed in the containing jar, the porous cell is to be put between the upright portions of the zinc, and the platinum plate is then put into the porous cell, *Fig* 29. The zinc plate is usually made out of a long stip bent up in the form of the letter U, by which means the zinc is brought opposite to each side of the platinum plate.

Fig. 29.

But it is more advantageous, instead of bending a long slip of zinc, to employ three shorter pieces. One piece to be put at the bottom of the containing jar; and two other pieces resting on this to form the two vertical sides. This is less expensive to make, and more economical in use. A binding screw attached to the zinc plate, and one to the platinum, form the two electrodes.

Two different exciting fluids are required for this battery—one for the platinum cell and the other for the zinc. That for the platinum is nitric acid (commercial aqua fortis) 3 parts, and sulphuric acid 1 part; and that for the zinc, sulphuric acid (commercial oil of vitriol) 1 part, by measure added to 7 parts of water. This battery is convenient in use, and is admirably adapted for general galvanic experiments, especially so for the intensity coil, but it is not generally suitable for manufacturing purposes.

30. Bunsen's Carbon Battery.—The elements of this battery are zinc for the *positive*, and carbon for the *negative* element. A porous cell and containing jar, and two exciting fluids, are required for this battery. The general arrangement of the Bunsen's Battery, and the nature of the exciting fluids, are precisely similar to those of Grove's Battery (**29**). The carbon usually employed for Bunsen's Battery is obtained from the deposited carbon that is found in the retort at the gas works.

31. Callan's or Maynooth Battery.—A cast-iron vessel is used in this battery as the containing cell, and this forms the *negative* element. A porous cell is placed within the iron one, and the zinc plate, which forms the *positive* element, is put into it, *Fig.* 31. Nitric acid is used in the iron cell as the exciting fluid, and sulphuric acid (commercial oil of vitriol), 1 part, by measure, added to 7 parts of water, is used in the porous cell. Many experimenters admire this battery; it is moderately cheap to construct, exposes a large negative surface, and evolves a powerful current; but its cost in use is considerable, owing to the large quantity of nitric acid required.

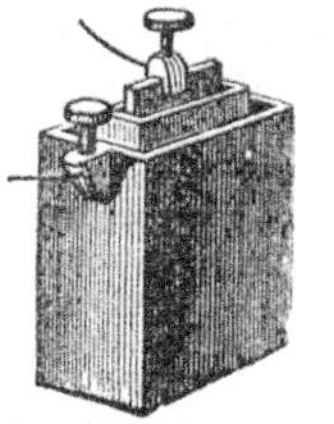

Fig. 31.

32. Bichromate Battery.—This consists of a plate of gas carbon for the *negative* element, and a plate of zinc for the *positive*. These plates are attached to a bar, and suspended in the containing jar; no porous cell is required, and one exciting fluid only is used. The exciting fluid is thus made:—Dissolve powered bichromate of potassa in hot water, the proportion being about 2 ounces of bichromate to 1 pint of water. When cold, add sulphuric acid (commercial oil of vitriol) in the proportion of 1 part of acid to 12 parts of bichromate solution. The addition of the acid will cause the evolution of a considerable degree of heat, and when the solution is again cold it is fit for use. This battery is convenient in use, does not emit any fumes, is very energetic, and will maintain its power about three hours.

A very convenient preparation has recently been introduced for forming the solution for the Bichromate Battery. It is known as "Crystals for the Bichromate Battery," and it only requires to be dissolved in the proportion of 1 oz. of the crystals to 4 oz. of water, and the solution is ready for use, without any further addition.

A useful form of the Bichromate Battery is that shown in *Fig.* 32, which has a glass bottle for the containing cell, and the zinc plate is in the centre, with a carbon plate on each side. The zinc is attached to a rod that passes up though the cap of the bottle, so that it can be drawn up or down. In charging the bottle-shaped cell with the exciting fluid, the vessel should not be filled higher than the body, so as to leave the neck empty. The zinc plate is made somewhat shorter than the neck of the bottle, and when the battery is not required for use the zinc can be drawn out of the solution in the body into the empty neck of the bottle, and then all action at once ceases. The solution can be left in the cell, and will keep good any length of time, and the battery can be put into action whenever required.

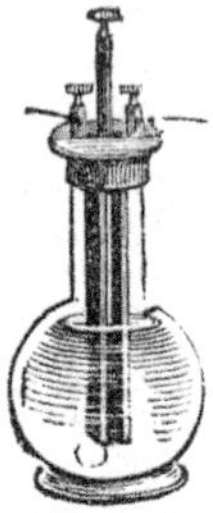

Fig. 32.

All the batteries described (**27** to **32**) can be used either as single or compound batteries, and can, if arranged as compound batteries, be connected either for quantity or intensity effect.

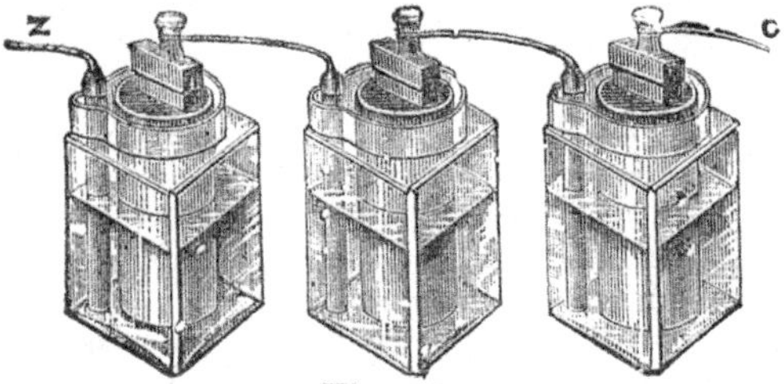

Fig. 32a.

32*a*. Leclanché Battery.—This form of battery, *Fig.* 32*a*, is in very general use for Electric bells, its great recommendation being that once charged, it retains its power without attention for several years.

There are two jars employed in the construction of the Leclanché Battery ; the outer one is of glass, and contains a zinc rod, and is charged with a solution of chloride of ammonium (sal ammoniac). The inner jar is of porous earthenware, and contains a carbon plate, and is filled up with a mixture of peroxide of manganese and broken gas carbon.

When the carbon plate and the zinc rod are connected, a steady current of electricity is set up ; the chemical reaction which takes place being as follows :—The zinc becomes oxidised by the oxygen from the peroxide of manganese, and is subsequently converted into chloride of zinc by the action of the sal-ammoniac.

After the battery has been in continuous use for some hours, the manganese becomes exhausted of oxygen, and the force of the electrical current is greatly diminished; but, if the battery be allowed to rest for a short time, the manganese obtains a fresh supply of oxygen from the atmosphere, and is again fit for use.

After about eighteen months' use, the glass cell will probably require recharging with sal ammoniac, and the zinc rod may also need renewing; but, should the porous cell get out of order, it is better to get a new one entirely than to attempt to recharge it.

33. Amalgamation of Zinc Plates.—The Zinc plates used for the Smee's, Grove's, Callan's, and Bunsen's Batteries should be amalgamated when the battery is first made, and they will require to be re-amalgamated from time to time. This is very readily effected by laying the zinc plates in dilute sulphuric acid of the strength used for the exciting solution, and allowing the plate to remain for some minutes. It should then be taken out, and, while wet, have a globule or two of mercury placed on it, and rubbed over its face and edges, so that it may present a uniform silvered appearance. This amalgamating is necessary in order to prevent what is technically called *local action*, and its necessity may be thus exemplified. If a piece of *unamalgamated* zinc be put into dilute sulphuric acid, the surface will be immediately covered with minute bubbles of hydrogen gas, and the acid solution will soon be in a state of effervescence, accompanied with a hissing sound. This will go on until the zinc is entirely dissolved, or the solution completely saturated with sulphate of zinc. But if the zinc plate be amalgamated, no action of this kind will take place; the smoothness of the surface communicated by the mercury appears to cause the hydrogen first evoluted to *adhere* to the plate, which is thus protected from any further action. This immunity only exists while the zinc plate is detached from the other part of the battery, or the circuit between the anode and the cathode is severed. When the zinc plate is in its place, and the circuit completed, the mercury affords no protection, as the hydrogen then is evolved at the negative element (**3**). When the zinc plates of a battery, after being washed, look black in parts, it is needful to re-amalgamate them previous to using them again.

34. Elements of a Battery to be kept apart.—It will be obvious, from the foregoing remarks, that the positive and negative elements of a battery must not touch each other *within* the exciting fluid. They should be about $\frac{3}{8}$ to $\frac{1}{2}$ inch apart, and the batteries that are used without porous cells will require to be examined from time to time, in order to see that these conditions have not become deranged.

INDUCED ELECTRICITY.

35. Induced Currents.—In a previous experiment (**8**) it was shown that an electrified wire is capable of exciting an electrical current in a wire that is contiguous to but not connected with it. This kind of current is called an Induced Current. The Induced Current differs strikingly from the Battery Current, inasmuch as the Battery Current is continuous as long as the circuit is complete, whereas the Induced Current is but momentary, and requires a disruption of the Battery Circuit to re-excite it. The experiments (**36** to **43**) will serve to illustrate the phenomena of Induced Currents, and will render more obvious the descriptions that will be given.

36. Induced Current on Completing Circuit.—Fix two wooden pillars, A B, into the top of a wooden base, *Fig.* 36. The height of these pillars, or their distance apart, is not of material moment. If the pillars be made 6 inches high, and be about 12 inches apart, such a proportion will be convenient. The top of each pillar should have two binding screws in it, so that the bent wires, C D and E F, may be fitted in one below the other, and about ¼ inch apart. The Galvanometer, G, is to be connected with the two binding screws carrying the lower wire, E F, so as to form a continuous and closed circuit. The cathode of the Battery, H, is now to be connected with the binding screw, D, forming one of the ends of the wire, C D. It will be seen that the battery circuit is not yet complete, for the wire from the binding screw, C, is not connected with the anode of the battery; therefore all is quiescent. If the battery circuit be completed by touching the anode of the battery with the wire from the binding screw, C, the needle of the Galvanometer will be immediately deflected into an oblique position. This deflection, as observed before (**8**), is caused by the induced current of electricity that is set up in the *galvanometer* circuit by the action of the current that circulates through the *battery* circuit as soon as this latter is completed.

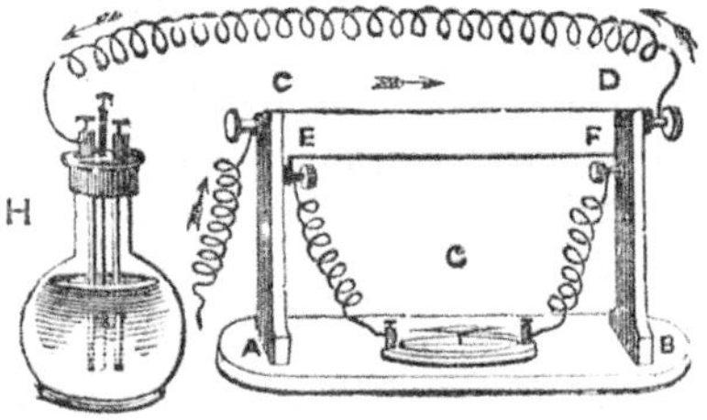

Fig. 36.

37. Induced Current on Rupture of Circuit.—Let this be pursued further. Instead of simply *touching* the anode by the wire, C, let the wire be *fixed* to the anode by means of its binding screw. A similar deflection will take place, but it will be seen that it exists only for an instant; the needle will soon return to its normal position, although the contact of the wire with the battery is still preserved. This shows that an induced current differs from the battery current in that it is but transient (**35**), and does not flow continuously. Separate the wire, C, from the anode of the battery, and again the needle will be deflected; but it will be observed that the deflection is now in the *opposite* direction to the former deflection. This, also, only lasts for a moment, and the needle will then return again to its original position.

38. Necessity for Rupture of the Battery Circuit.—These experiments will enable the phenomena of induced currents to be better understood, for they show that the influence of an electrified wire upon a contiguous wire brings about an alteration in the electrical state of such wire; and this, in the present case, is shown by the deflection of the needle of the galvanometer. It is also seen that this electrical manifestation exists only *while* the alteration in the electrical condition of the wire is being brought about, for, as soon as the alteration is effected, it ceases. But more than this is shown, as it is manifest that the battery current has the power, not only of *producing* this alteration, in the electrical state of the contiguous wire, but also of *maintaining* that altered condition, for the needle soon returned to its normal condition, although the battery current continued to flow. Even this is not all; for when the battery circuit was interrupted by separating the wire from the anode of the battery, then the force that had been operating ceased, and the wire in the galvanometer circuit returned to its normal condition; but in doing this another current was set up, manifested by the second deflection of the needle, a current, be it observed, differing in some respects from the first, inasmuch as the deflection of the needle was in the opposite direction.

39. Rapid Rupture of Battery Circuit.—Let the mode of performing the experiment (**36**) be varied by attaching the wire from the file, J, *Fig.* 39, to the anode of the battery. The file under these circumstances becomes the electrode; and if the end of the wire from C be brought into contact with the file, the needle will be deflected as before. Make contact by again touching the file with the end of the wire from C, and draw it along the surface of the file. The proceeds will give a succession of contact-making and contact-breaking; for every

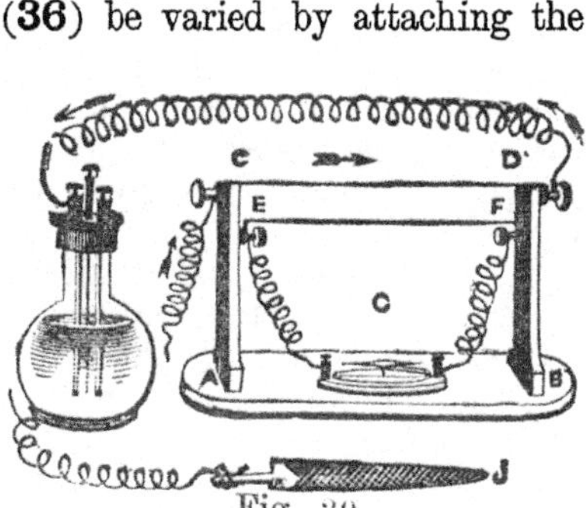

Fig. 39.

time the wire passes off one tooth of the file, contact will be broken, and each time it again touches the next tooth, contact will be renewed. Under these circumstances, the needle of the galvanometer will be kept in a state of oscillation. The rapidity of the operation of making and breaking contact enables the induced electricity, which only exists for a moment, to approximate to the character of a continuous current.

40. Induction in the Battery Circuit.—There is another condition of induction that will require to be noticed in order to enable all the phenomena of the Intensity Coil to be understood. In the preceding experiment the action of an electrified wire upon an independent wire was noticed; it will now be seen that, under certain circumstances, an electrified wire acts upon *itself*. Prepare a battery (say that shown at *Fig.* 32), and attach one end of the insulated wire, about 18 inches long, to the cathode. The other end of the wire is to be brought into contact with the anode, and then separated, when a *minute* spark will be seen at the point of rupture, *Fig.* 40.

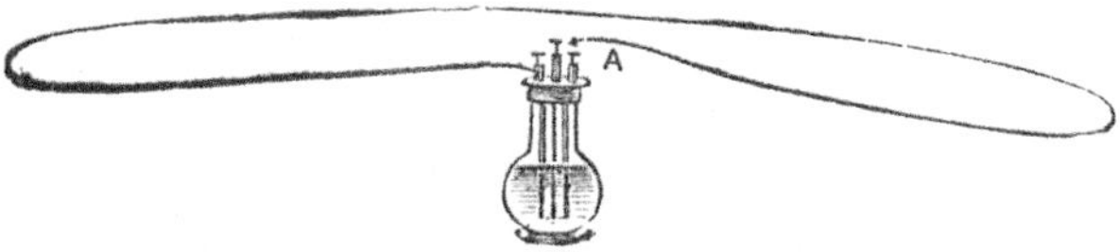

Fig. 40.

41. Induction in Scalariform Battery Wire.—Remove the wire from the battery, and bend it into a scalariform or ladder-like shape. Repeat the former experiment with the wire thus arranged, and the spark will be perceptibly larger and brighter.

42. Induction of Spiral Battery Wire.—Change now the form of the wire by winding it into a spiral or helix, C, *Fig.* 42. On repeating the experiment with the wire in this shape, the spark is still further increased.

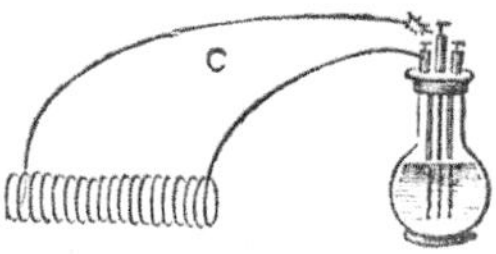

Fig. 42.

43. Magnetic Induction in Battery Circuit.—Repeat the former experiment (**42**), first inserting the iron rod D, *Fig.* 43, into the centre of helix, C, and when contact with the battery is broken under these altered circumstances, the force and brilliancy of the spark will be further intensified.

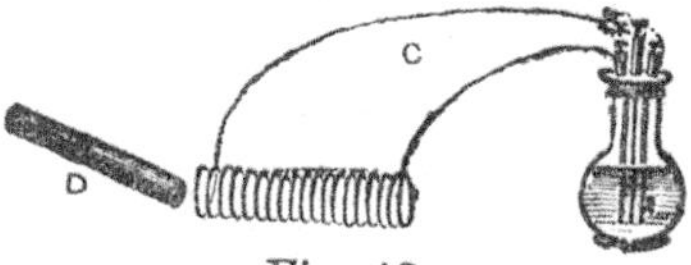

Fig. 43.

The explanation is this : In the first experiment (**40**) the spark was produced by the combustion of a portion of the conducting wire by the unassisted force of the battery current. In the second and third experiments (**41** to **42**) the *increase* in the spark was due to increased combustion of the conducting wire, arising from the inductive action of one fold of the wire upon that contiguous to it. In the fourth experiment (**43**) the iron rod in the centre of the helix was rendered magnetic by the flow of the battery current round it, and this magnetised rod, in its turn, acted on the helix, and thus still further increased the force of the current.

If a galvanometer (**7**) be made a part of the circuit in performing these experiments (**40** to **43**), the effects described will be seen in the increasing length of the arc through which the needle will be deflected. It must be noticed that these increased effects are manifested at the moment of breaking battery contact, and, like the former results (**36**), they are but transient, and only approximate to continuity by a rapid process of contact breaking.

44. Electro-Chemical Effects of Current Electricity.—Prepare a piece of white blotting paper by immersing it in a solution of starch, and then drying it. Moisten the paper thus prepared with a solution of iodide of potassium, and lay it on a glass or ebonite plate. Arrange the battery so that the wires (which in this case must be platinum) from the electrodes may come down upon the moistened paper. *Fig.* 44. Let the ends of the wires be about ½ an inch apart, and in the course of a few seconds there will be a violet spot under the end of the wire from the anode of the battery. This colour results from the decomposition of the iodide of potassium by the galvanic current and the union of the iodine thus liberated with the starch that the paper was prepared with, forming iodide of starch, which is of a deep violet colour.

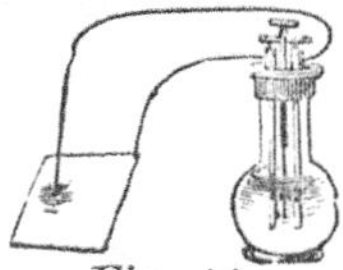

Fig. 44.

HOW A COIL IS MADE.

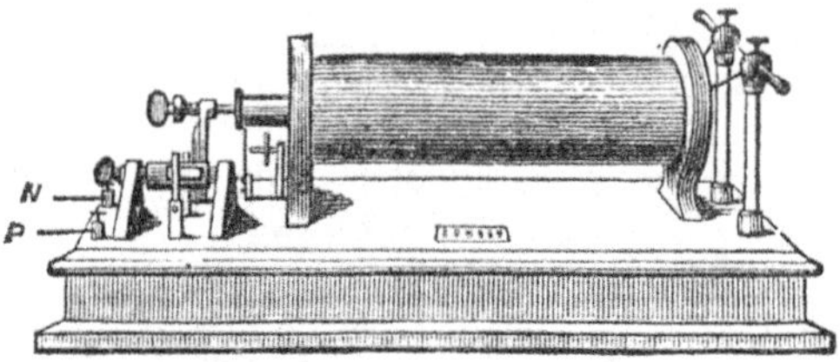

45. The Primary Coil—From experiments such as the preceding it is obvious that by arranging the wire used for uniting the two electrodes of a galvanic battery into the form of a helix, certain properties of the battery are greatly increased. When a long length of wire is used, it is best to wind it in two or more layers, one over the other. Such a helix is designated a Coil, and, because it is employed for conveying the battery current, it is called a Primary Coil. Whenever the word Primary is used with reference to galvanic arrangements or experiments, it is intended to signify some relation to a battery circuit.

46. Construction of an Ordinary Primary Coil.—Prepare a paper tube, about 4 inches long and $\frac{3}{4}$ inch diameter, and wind on this tube two or three strands or layers of wire, one over the other. The wire should be copper, covered with cotton, and of the size of ordinary bell wire. A binding screw should be attached to each end of the wire, as shown in *Fig.* 46, by which means it can be conveniently united to a battery. This is a Primary Coil, and, by its aid, the results observed in **42, 43**, can be obtained in a more marked and palpable manner. When the paper tube upon which the wire is wound is filled up with an iron bar, greatly increased effects are manifested, and if, instead of a solid iron bar, a bundle of iron wire, A, *Fig.* 46, be employed for this purpose, the improvement is most striking.

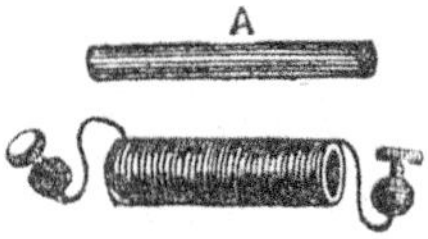

Fig. 46.

47. Construction of an Ordinary Secondary Coil.—Prepare a second paper tube of similar length to that in the centre of the Primary Coil just described, but large enough in diameter to fit over it. Fit two discs of wood on the ends of this tube, and then wind on five or six strands of cotton-covered wire about the size of stout packthread, and attach binding screws to the beginning and end of such wire. This, when placed over the Primary Coil, is termed a Secondary Coil, *Fig.* 47. It is con-

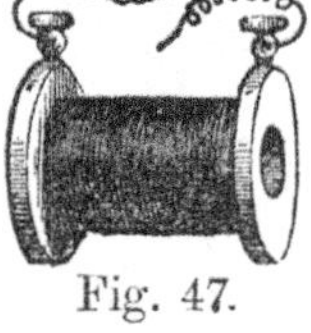
Fig. 47.

tiguous to the Primary Coil, but has no connection with it.

48. Sparks from Secondary Coil.—Place two thin wires in the binding screws of the Secondary Coil, and bring their ends near together, so that they may all but touch, and place the Primary Coil, with its core of iron wires, within the Secondary Coil. Attach the Primary Coil to a battery, and break contact rapidly by means of a file (**39**), when a series of minute sparks will be seen between the ends of the wires from the Secondary Coil (**47**). If the fingers be wetted and brought into contact with the two binding screws on the Secondary Coil, or with metal cylinders attached to them, a distinct shock will be felt when the contact between the Primary Coil and the Battery is made or broken. It is thus that the coils of the electro-magnetic machines are constructed that have been made and sold by thousands, both for medical and experimental purposes. But such machines are not Intensity Coils, for their mode of construction involves the loss of all but a mere fragment of the electrical current excited in the wires of which they are composed.

49. Contact Breakers.—As the induced currents (**36**, **38**) are only manifested when contact between the Primary Coil and the battery is made and broken, it will not be out of place to describe here the means by which this rupture is usually effected. The contrivance for this purpose is called a Contact Breaker, and it will only be necessary to describe two kinds, as they may be taken as types of those in ordinary use.

50. The Mercury Contact Breakers.—In *Fig.* 50 is represented the apparatus devised by Dr. Ritchie as a mode of obtaining rotary motion by the temporary magnetization of an iron bar, and this has been extensively employed as a Contact Breaker. It consists of a circular disc of wood placed between the poles of a horseshoe magnet, having a deep channel turned in it, so as to form a cup. This cup is divided into two parts or semi-cups, by a wooden bridge, the ends of which come opposite to the poles of the magnet. A brass pillar rises from the centre of the bridge, supporting on its top an iron bar wound round with insulated wire, the two ends of which come down into the cup, and are of such a length that when the iron bar is rotated, they will just pass over the bridge without touching it. This bar, or electro-magnet (**11**), as it really is, has a pointed pin projecting from its under side, which fits into the brass pillar, and this arrangement allows the bar to rotate with very little impediment from friction. The two semi-cups are to be filled with mercury, which will stand up above the top of the bridge, the bridge thus causing a sort of trough between them.

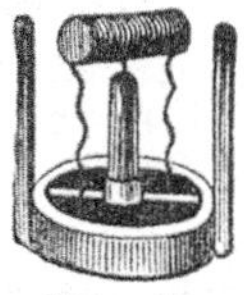

Fig. 50.

The wires from the electrodes of a battery are to be put into the mercury in the semi-cups, (one into each), and the rotating bar moved round, so that it may stand *across* instead of in the line of the bridge. As soon as this is done the wires from the iron bar will touch the mercury, and the battery current will then circulate round the bar, and convert it into an electro-magnet (**11**). The north and south poles of the horse-shoe magnet will attract dissimilar poles, produced in the iron bar by the action of the battery current in the iron bar, and draw them round until they are opposite to the two poles of the horse-shoe magnet. This operation will also carry the wires out of the mercury, and communication with the battery will be interrupted, and consequently the electro-magnet will lose all its properties.

At first sight it would appear that the electro-magnet would now come to a state of rest, but this is not the case; the impetus it acquired by its partial rotation will carry it beyond the line of the bridge, and this will bring the points of the wire again into the mercury in the semi-cups. But it will be noticed that the wires do not come into the same semi-cups as before, and that the battery current therefore flows through the wire on to the iron bar in the opposite direction; consequently the polarity now acquired by the bar is opposite to that which it had before. The end of the electro-magnetised bar that is *now* north is thus near to the north pole of the horse-shoe magnet, and these two mutually repel each other, and by this force the rotating bar is driven to a position at right angles to the bridge, and where its north can be attracted by the south of the horse-shoe magnet. This attractive force carries the bar through another quarter of the circle, and it is brought again, with its wires, over the bridge. But the wires coming again into the opposite semi-cups, opposite polarity is induced; again it is driven through a quarter circle, attracted through another quarter of the circle, until it is once more over the bridge.

In this way, by this alternative magnetization and demagnetization, an attractive and repulsive action is obtained, by means of which a rapid rotation is produced, and a contact made and broken twice in each revolution.

This, though a convenient Contact Breaker for some purposes, is not suitable for use when large batteries or coils are employed. Every time the wires leave the mercury, a vivid spark occurs, and the surface of the mercury soon becomes covered with a coating of oxide of mercury. This oxide, being a non-conductor, prevents the battery current from flowing into the wire, and so interrupts the action. These inconveniences are remedied by the form of Contact Breaker now to be described, and which is usually called the Vibrating Contact Breaker.

51. Vibrating Contact Breaker.—By reference to the illustration, *Fig.* 51, the general form of this Contact Breaker will be seen. It consists of a base board, having an outer brass pillar, a centre brass pillar, and an electro-magnet (**11**). The electro-magnet is fixed to the board with its poles upwards, and the ends of the wire that is wound round it are thus disposed of. One end is left open, so that the battery may be connected with it, and the other end passes under the board to the base of the centre pillar. The outer pillar, at its upper part, holds the end of a metal spring, which passes through the ring of the centre pillar to the poles of the electro-magnet. Here the end of the spring is armed with a plate or clapper of iron, which should stand, when the spring is at rest, about 1-10th of an inch above the poles of the electro-magnet. A screw passes through the ring at the top of the centre pillar, and comes just in contact with the spring. The spring at this point, as well as at the end of the screw, must be of platinum.

Fig. 51.

When one electrode of a battery is attached to the open end of the wire of the electro-magnet and the other electrode to the outer brass pillar, the circuit is complete. If the anode of the battery (**19**) be connected with the electro magnet, the current will enter there, circulate round it, communicate magnetic properties to it, and pass under the board to the centre pillar. It will rise up here to the ring, and descend through the screw to the spring, and from thence by the outer pillar to the cathode of the battery.

The electro-magnet will now attract the iron clapper at the end of the spring down to itself, and by this means a separation takes place between the end of the screw and the spring, and the battery circuit is interrupted. The electro-magnet then loses all its attractive power; it can no longer hold the clapper down, and the spring thus liberated rises to the position it formerly occupied, and again comes into contact with the end of the screw that passes through the ring. As soon as this takes place, the current again flows, the electro-magnet exercises attractive influence on the clapper and draws it down. This once more breaks contact, again the spring resumes its position, and thus a rapid vibration is kept up; every oscillation of the spring being associated with making and breaking contact with the battery.

When this Contact Breaker is applied to Intensity Coils, it is usual to employ the iron bundle forming the core of the coil (**46**) as the electro-magnet, and to place the vibrating spring vertical instead of horizontal.

HOW AN INTENSITY COIL IS MADE.

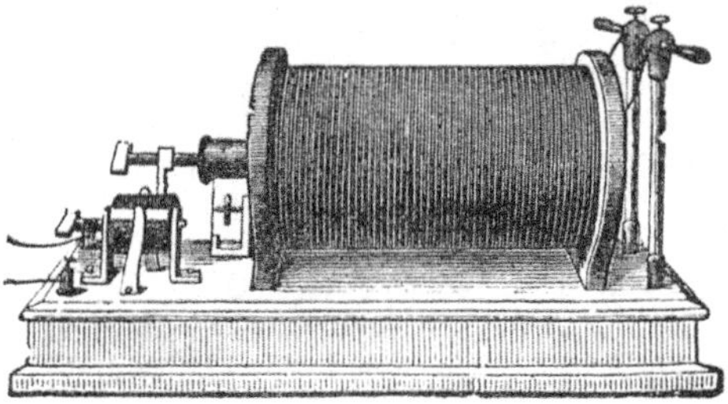

Fig. 52.

52. Intensity Coils.—It is to M. Ruhmkorff, of Paris, that the scientific world is indebted for a knowledge of the means by which the entire results of the combined action of battery, wires, and iron core can be obtained. M. Ruhmkorff, may, therefore, be properly regarded as the inventor of the Intensity or Induction Coil. As it is proposed to give the detailed instructions for making as well as using these coils, it will not be necessary to do more than advert thus to the inventor, as the elements of the invention itself will be found in the directions that are to follow.

The parts of an Intensity Coil are these: The Reel, the Primary Coil, the Secondary Coil, the Iron Bundle or Core, the Contact Breaker, the Condenser, the Pedestal or Base, and the Commutator. The dimensions about to be given are not intended to be taken as expressing the absolute relation that must exist between the various parts of an Intensity Coil. They may be considerably varied without impairing the efficiency of the apparatus.

53. The Reel.—The reel consists of a hollow cylinder or tube, with a square or circular plate firmly fixed on each end of the same. The cylinder should be formed of paper, and the plates or Reel ends, as they may be called, should be of gutta-percha or ebonite.

The Reel ends should be flat, and not less than $\frac{3}{8}$ inch thick, and if circular, a facet should be made on the edge of each, so that when the Reel is complete it may stand steadily on the pedestal. The hole through the centre of the Reel ends should be turned perfectly true, so as to fit the outside of the cylinder, and a shoulder should be left on the outer face sufficient to prevent the paper cylinder from being pushed through the ends when it is being fastened on. The Reel ends may be $4\frac{1}{2}$ inches diameter if circular, or 4 inches by

4 inches if square. They can be fastened to the paper cylinder by Thompson's Liverpool glue or cement.

The cylinder should be formed of cartridge paper cut into a long strip, and when gummed or pasted on one side, wound round a rod 7-8ths inch diameter. When properly done, a firm tube, 7 inches long, 1 inch diameter, and about 1-16th of an inch thick is obtained. This is to be allowed to dry thoroughly, and the ends cut perfectly at right angles to the axes. The discs or ends must be firmly fastened to the cylinder, and in order to effect this the holes through the discs should be slightly tapered, the larger dimensions being towards the shoulder. Before fixing the paper cylinder, a slightly conical plug should be provided, fitting the inside of the cylinder. When the parts are ready, and carefully coated with the glue, they are to be put together, and the conical plug gradually pressed into the end of the cylinder, which will expand it a little, and force it into close contact with the sides of the hole in the disc, and it is to remain in this position until the glue is thoroughly set, when the plug can be removed. The reel should be provided with a hollow groove in the edges of the discs, if they be circular, in order to receive the pieces of cat-gut cord that are to fasten it to the base; if they be square they can be fastened by screws. Two holes are to be drilled through one of the ends of the reel, in order to allow the primary wire to be passed through. These holes should be about $\frac{1}{8}$ inch diameter, and they should be somewhat oblique in their direction, so that the wire, where passed through the reel end, may not be at right angles with the axes of the reel.

54. The Primary Coil.—The Primary Coil consists of copper wire covered with cotton, and of a size known as No. 16, averaging about 18 yards to the lb. One end of the wire is to be passed from the inside through one of the holes in the reel end, so as to project six or eight inches, and the wire is then to be carefully wound over the cylinder up to the other end, and then back again, so as to form two layers of wire, one over the other. When this is completed, the remaining end of the wire is to be passed through the second hole in the reel end. Such a quantity of wire will weigh about one lb. Before putting on the wire it is necessary to fit a wooden or metal rod inside the paper cylinder of the reel. If this be not done, the cylinder is likely to be damaged by the force required to wind the wire round it.

When the primary wire is laid on, it should be varnished with two or three successive coats of varnish, composed of shellac dissolved in spirits of wine. Care should be taken that one coat be thoroughly dry before a second is put on. The varnish for the first coat should be thin, so that it may be readily absorbed by the covering of the wire, and conveyed to the underneath layer. When the varnish is dry and hard, the primary wire is to be carefully covered with a strip of cartridge paper passed twice or three times over the wire, and fastened by gum-water or glue. This paper must be cut exactly to the width between the inside faces of the ends of the reel. It must be drawn tightly when put on, but not so tightly as to show on its surface the interstices between the rows of wires. This paper covering, when dry, is to be carefully varnished, and must present a smooth cylindrical surface, having no interspace between it and the inside face of the reel ends.

Shellac varnish, laid on as above described, forms a good insulator, but it is not quite so effective a mixture as the ordinary black resin of commerce and beeswax. This preparation is rather more difficult to apply, but it is greatly superior to the varnish when it is done.

The method to be adopted is this: the resin is to be melted in an earthen vessel, and a small quantity of beeswax added to it; the proportion of wax has to be determined by experiment, the use of the wax being to diminish the friability of the resin, without interfering with its hardness. Usually, about one-fifth by weight of wax will be found a suitable quantity. The resin and wax, when fully melted and heated almost to boiling, are to be carefully poured over the wire by means of a spoon or ladle, turning the coil round and repeating the application until the mixture has completely permeated the strands and filled up all the interstices between the wires.

If this be done neatly, the paper covering will not be required; should, however, the outer surface of the coil be otherwise than smooth and even, the paper must be laid on as directed.

When the resin mixture is employed as the insulating material, it will be found convenient to wind the wires on to the cylinder, and insulate the same, before the reel ends are fixed on.

55. The Secondary Coil.—This is also formed of copper wire, but the wire must be covered with *silk*, and of a size known as No. 38, averaging about 180 yards to the ounce; the quantity required will be about six ounces. In winding this wire on the reel, the most scrupulous care is needed, not only to avoid any break in the wire, but also any kink or bend in it. The diameter of this wire is ·0067 inch. The layers of the Secondary Wire should not be carried close up to the end of the primary coil, in order to avoid the possibility of the wire of one layer sinking down to a level of that below it. It must be borne in mind that when a layer of wire is finished and insulated, it is next to impossible to take it off again; consequently every care must be taken to prevent any failure in winding it on. One layer of the Secondary Wire is to be wound on at one time, and then most carefully coated with the shellac varnish or with the resin mixture (**54**). When this has been done, the layer of wire is further insulated by wrapping round it several thicknesses of gutta-percha tissue, or of thin white demy paper, previously soaked in the resin mixture (**54**), and allowed to become hard.

Whichever material be used, it must be cut in strips a little wider than the length of the layer of wire it is to cover, and then wound on tightly, but smoothly. The length of the strip should be such as to wrap three or four times round the coil, and must be fastened with the varnish or resin mixture. When the requisite quantity of wire is put on, eight or ten folds of the insulating paper or tissue must be wrapped round the coil before the ornamental covering of silk velvet is applied.

If the wires forming the coil have been put on before the reel ends have been fixed to the inner paper cylinder, the reel ends must now be put in their places, and when they are firmly set, the spaces between the ends of the layers of wire and the inside of the reel ends should be carefully filled up with the resin mixture (**54**) so that the insulation may be perfect.

The winding of the Secondary Wire should begin at the *opposite* end of the coil to that at which the winding of the primary coil commenced, and it must finish at the same end at which it began. The two ends of the wire should be wound into helices, and these can be passed through two holes in the reel ends, in order that they may be connected with the other part of the apparatus.

56. Testing the Secondary Wire.—Before the winding of the Secondary Wire is commenced, it should be tested, which may be done in the following way:—Attach one end of the wire on the bobbin (as it comes from the covers) to one electrode of the battery, and the other end to one of the binding screws of a galvanometer (**7**). The circuit is now to be completed by uniting the other electrode of the battery with the other binding screw of the galvanometer, and if there be no break in the wire a deflection of needle will ensue. Should *no* deflection take place, the wire must be unwound from the bobbin and carefully examined and the break detected and soldered.

When the continuity of the wire has been ascertained or effected, the operation of winding it on to the reel can be commenced. Each layer of wire, as it is wound on, and before it is insulated, should be tested by the galvanometer. For this operation a different course to that just described should be adopted. The beginning of the Secondary Wire on the reel is to be connected with one of the binding screws of the galvanometer, and the end of the wire that still remains on the bobbin is to be attached to the other binding screw of the same, *Fig.* 56. The two ends of the primary coil are to be connected with the battery, and if there be no fault in the Secondary Wire, the needle of the galvanometer will be deflected; the layer can then be insulated (**55**), and the like course must be adopted with each layer.

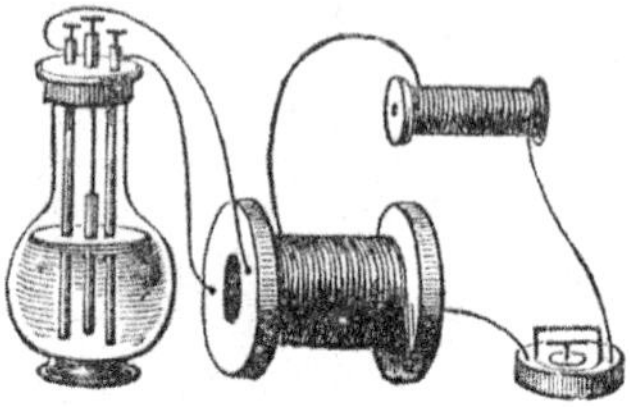

Fig. 56.

Should it be necessary to solder two ends of the Secondary Wire together, the silk coating must be removed from each end. The ends must then be carefully brightened by rubbing them with fine glass paper. The ends of the two wires are to be put side by side in contact with each other, and a small piece of tin-foil wrapped round both wires, and moistened with a solution of muriate of zinc, and then moved over the flame of a very small spirit lamp, and in a few seconds the tin-foil will melt and unite the wires together, Should it be necessary to apply the wires again to the lamp, they must be first moistened with a fresh portion of muriate of zinc. The wires should overlap each other about half an inch, and when the soldering is complete the silk covering is to be most carefully replaced.

57. The Iron Bundle.—This is simply a bundle of uncovered iron wires, about No. 18 gauge. The wires should be quite straight, and of exactly equal lengths, and about $\frac{1}{2}$ an inch longer than the outside measurement of the coil. The centre of the coil is to be filled with these wires, and when this has been done, a short piece of larger wire, carrying on one of its ends a disc of iron, about $\frac{1}{4}$ inch thick and $\frac{5}{8}$ inch diameter, should be pushed into the centre of the coil at each end, so as to secure the bundle in its place.

58. The Contact Breaker.—The form of Contact Breaker used for Intensity Coils, is that known as the Vibrating Contact Breaker, a description of which has been previously given. In the form described (**51**) an electro-magnet was provided, so that such a Contact Breaker could be applied to any kind of apparatus whatever. But it is not desirable to use a separate electro-magnet for Intensity Coils, as a resistence is offered by it to the passage of the battery current, and therefore the iron bundle in the coil (**57**), which becomes an electro-magnet, is used instead.

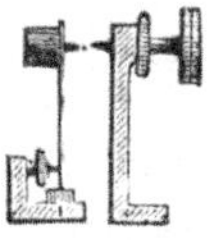

Fig. 58.

This arrangement necessitates an alteration in the position of the spring and iron clapper, which, as will be seen in the illustration *Fig*. 58, is placed vertically. The spring is fixed to a brass block attached to the pedestal, the block having a vertical plate rising on one side. A screw passes through this vertical plate of the block, and comes into contact with the spring a little way above its point of fixation to a block. The use of this screw is to increase or decrease the tension of the spring, and to regulate its distance from the ends of the iron bundle. At the top end of the spring is an iron cylinder or clapper about $\frac{1}{2}$ inch long, and of similar diameter, and the spring should be adjusted so that the face of this cylinder may, when the spring is at rest, be about $\frac{1}{4}$ inch from the end of the bundle. A strong brass pillar rises up also from the pedestal, and reaches a little above the centre of the coil. Through the top of this pillar a strong screw (called the platinum screw) passes, carrying on its end a piece of platinum, which comes into contact with the spring at or a little below where the iron cylinder is attached to it. The spring at this part is armed with platinum, and it is here that the contact is made and broken. The platinum screw is provided with a running boss, so that when the screw is adjusted the boss can be brought up tightly against the pillar, and thus prevent the screw from shifting. The surfaces of the platinum will require to be smoothed and scraped from time to time, in order to maintain complete contact. It will not be necessary to repeat here the reason why the spring vibrates; that has been fully described before (**51**).

59. The Condenser.—This part of the apparatus is usually shut up in the cavity of the pedestal forming the base, although it can, if preferred, be separate. The Condenser is the invention of M. Fiseau, of Paris, and its purport is to add to the energy of the current that traverses the primary wire, and consequently to increase the force of the secondary discharge. The Condenser consists of a number of plates of tin-foil, separated by sheets of carefully varnished or resinised (**55**) paper, the alternate tin-foil plates being attached together, thus forming two separate insulated series. One series of the plates is connected with the pillar of the contact breaker that carries the platinum screw, and the other series with the block that holds the vibrating spring. It will be seen from this that these plates do not form part of the battery circuit, but are, as it were, lateral expansions of that circuit, on each side of the contact breaker. The insulating sheets between the tin-foil plates have by these means their electrical condition disturbed, and when the battery circuit is interrupted the plates return to their normal state, and, in so doing, increase the action of the current circulating in the primary wire.

The paper intended to be used for separating the plates of the Condenser should be moderately thin, and not too heavily sized. The paper should be cut into pieces, rather larger than is required, and then dipped into a solution of 1 oz. of shellac dissolved in 6 ozs. of methylated spirit, and hung up to dry. The paper should remain suspended for some hours, and then be carefully examined sheet by sheet; and if the minutest pinhole be observed in any sheet it must be rejected. A second coating of shellac varnish is now to be applied, and when thoroughly dry, the paper is ready for use. It must then be cut to the proper size, and preserved in a portfolio until required for use. Resinised paper can be used instead of the varnished paper if preferred. In that case it should be prepared as before directed (**56, 57**). The ordinary tissue paper does well for the purpose, but that known as white demy is perhaps the better kind.

The details of a Condenser for the coil being described are these: Prepare 50 sheets of tin-foil, each 5 inches long and 5 inches wide, and also 60 pieces of insulating paper, each 7 × 5, and two thin boards of mahogany of rather smaller size, varnished on each side. One of the mahogany boards is to be laid down, and upon this five of the insulating papers are to be placed, and then one of the tin-foil plates, taking care to lay this latter down so that one inch of its breadth projects over one side of the varnished paper. Another paper is to be laid on this, coinciding in position with the first paper, and upon this the second tin-foil plate is to be laid, but with the overhanging part coming out at the opposite side to that of the former one. This is to be covered with an insulating paper, and then a third tin-foil plate is to be laid down, but the overhanging part of this plate is to be on the same side as the first plate. This is to be followed by the other plates in similar order, taking care that the ends of each alternate plate projects at one side, and the ends of the intervening ones on the opposite side. When this has been done, five more papers are to be laid on, and, finally, the second mahogany board, and the whole is then to be tied up with gutta-percha string. All the projecting tin-foils at one side are to be pressed together, also those at the other side; the Condenser is then ready to be placed in the cavity of the pedestal.

60. The Pedestal or Base.—This should be made about 13 inches long and 8 inches wide, and 2 inches deep. The bottom must be made movable, and should be fixed by screws or buttons, and any kind of moulding can be placed round the Pedestal to suit the taste of the constructor. The coil is to be placed horizontally in the centre of the Pedestal, and it will be necessary to make holes in the top in order to fix the coil in its proper position. Other holes will be required in the top of the Base to allow the ends of the primary wire, together with the pillars and binding screws, to be passed through, in order to attach them to their proper places underneath the same.

The contract breaker is to be fixed in its place at one end of the coil, and two binding screws are also required to be fitted to the same end of the Pedestal. At the other end of the Pedestal two pillars of ebonite are to be placed. These may be six inches high and about ½ inch diameter. If the ebonite be cut off about an inch longer than is required, the extra inch can be turned down to a pin about a ¼ inch in diameter, and a screw cut on its end. The holes in the Pedestal are to be made sufficiently large to allow these pins to pass through, and the pillars can then be firmly fixed by putting a nut on the under side. On the top of each pillar should be a double binding screw, that is, a binding screw with two holes, and a separate screw to each. One of these holes is for the reception of one of the ends of the secondary wire of the coil, and the other, a hole in the same binding screw, is to attach any apparatus that is to be employed in conjunction with the coil.

61. The Commutator.—This instrument is shown at *Fig.* 61, and its use is to change the direction in which the current flows through the primary circuit, and, of necessity, to change the direction of that in the secondary circuit also. This, though an extremely useful addition to an Intensity Coil, cannot be regarded as one of its essential elements.

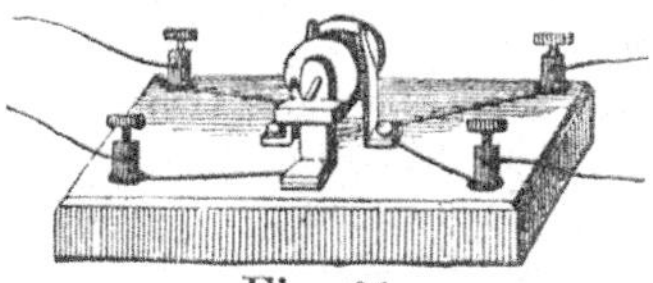

Fig. 61.

The Commutator consists of a cylinder of ivory or ebonite, and may be one inch long and one inch diameter. A metal axis should project from each end, and these two axes are not simply the end of one piece of metal running through the cylinder; they are separate pieces, and must not be united together. Two plates of brass, about ¾ inch wide, are fixed to opposite sides of the cylinder, one of these plates being connected with one axis, and the other with the other axis. The cylinder is supported horizontally, on two brass blocks or pillars that are fixed to the base board, and there are also two brass springs that rise up from the board and press on the brass plates on the face of the cylinder. Four binding screws are also on the boards; two are connected with the two springs by wires passing underneath or over the base board, and the other two binding screws with the blocks carrying the axes of the cylinder. One of the axes projects through the block in which it rests, and on this is fitted an ivory or ebonite plate, so as to enable the cylinder to be turned round when required. Two of the binding screws on the board are to be connected with the battery, and the other two binding screws with the apparatus to be operated with.

The current proceeds from the anode of the battery, through one of the binding screws of the Commutator, and from thence under or over the board to one of the springs, and proceeds up this to the plate on the ivory or ebonite cylinder it is in contact with. From thence it passes to the first axis, then through the block in which the axis rests, and out by the binding screw connected with that block to the apparatus, returning by the other spring plate and second axis to the battery. When the cylinder is turned half round, and without any other change in the arrangement, the current will pass through the *apparatus* in the opposite direction to what it did before. It will still proceed from the anode to the battery to the first spring, but that spring is now in contact with the plate that is connected with the second axis, and, therefore, the current will proceed from this axis, and return to the battery by the first axis.

If the pedestal on which the coil is mounted be made sufficiently large, the Commutator can be fixed on to it, and then the separate stand and binding screws will not be required. The Commutator is not only capable of being used as a current reverser, but also as a current suspender; for if the cylinder be turned only one quarter of a revolution, the springs will rest upon the interspace between the two brass plates, and contact will be broken. This should always be done while the arrangements for the secondary current are being made, in order to avoid receiving what might prove an unpleasant shock.

62. The Coil Connections.—A word of explanation may be here given as to the way of making the connections underneath the base board, and it will be assumed that the commutator is not on the base board, but will be, if required to be used, a separate instrument. On turning the pedestal of the coil upside down, the under side will, when all the parts are connected together, present the appearance shown in *Fig.* 62. The letters A and B represent the ends of the binding screws to which the battery is attached; C is the end of the pillar of the contact breaker that carries the platinum screw; D is the end of the block that carries the spring forming the other part of the contact breaker; E and F are the beginning and end of the primary coil.

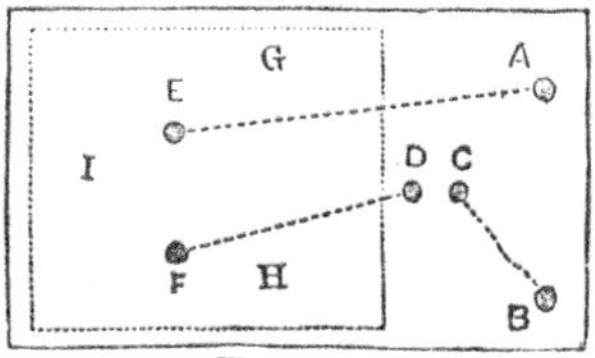

Fig. 62.

A loose board, I, should be fitted into the pedestal, and it should be fixed about half-way up from the bottom. This is required to lay the condenser on, and to afford convenient means for attaching it. The dotted lines represent this loose board, and upon it at G and H, two flat plates of brass, about 1 inch square, are to be fixed, having a screwed pin in the centre of each. These plates are for connecting the condenser with the contact breaker. The various screws spoken of should project through the top of the pedestal full $\frac{3}{8}$ of an inch, so as to enable nuts to be screwed over them. Three strips of sheet copper, about $\frac{1}{2}$ inch wide, are to be prepared, two of them having a hole at one end and a binding screw at the other, and the third is to have a hole at each end. This latter one is to be long enough to connect B with C, as shown by the dotted line. This is to be done by dropping the strip over the projecting screws of B and C, then screwing a nut down tightly on to the copper, which will secure the strip, and make the requisite contact. The strips, with binding screws, are to be used to connect A with E, and D with F, the binding screws securing the wires E and F, and the holes in the copper strips enabling them to pass over the projecting screws D and A, where they are fixed by nuts, as in the former case. When the anode of the battery is attached to the binding screw which is seen at A, and the cathode to that at B, then the current will flow through the coil. The circuit will be seen to be complete in the direction of the dotted line, A to E, through the primary coil, and out at F, then from F to D, and through the contact breaker to C, and from C to B.

The condenser is to be laid down on the loose board, I, with the projecting tin-foils resting on the brass plates, G and H. A hole may be punched through each set of tin-foils to admit the screw pin passing through them. A brass plate can then be laid on the top of the tin-foils, and by means of a nut on the screw pin they will be pressed closely together When the condenser has been fitted in its place, the nuts are to be loosened and the top brass plate taken off, and two copper strips with holes at each end fitted over the pins. These copper strips must be long enough to connect the pin, G, of the condenser with, C, of the contact breaker, and the pin, H, with D, and when this is done the nut may be screwed up tightly again. Should the condenser not be thick enough to fill up the space between the loose board and the bottom, the latter must be padded, so that when it is put into its place it will press tightly on the condenser, and prevent it from shifting.

63. Battery Power for the Coil.—The Coil is now complete, and it may be used with any battery equal in power to Grove's Battery of six cells. Less battery power can be employed if so desired, but then the full effects of the entire combination will not be obtained, and if the battery power be greatly in excess, the secondary wire may become heated, its insulations impaired, and the Coil utterly ruined.

HOW AN INTENSITY COIL IS USED.

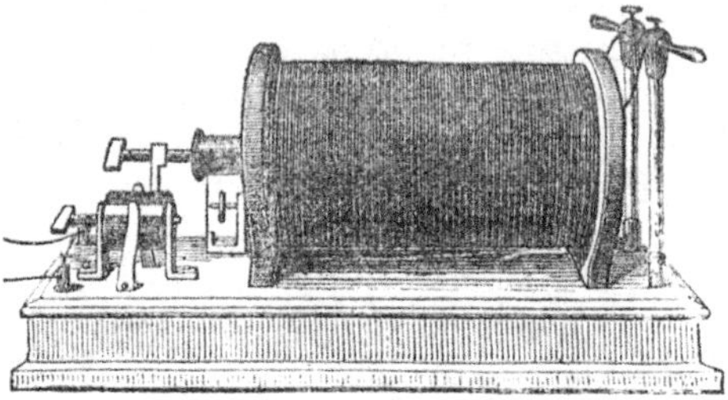

Fig. 64.

64. Electrodes of the Secondary Coil.—If two wires be affixed to the binding screws that represent the ends of the Secondary Coil, these wires are designated the Electrodes (**19**) of the Secondary Coil. And if these wires be attached to the discharger mentioned below (**66**), then the ends of the rods of the discharger are called the Electrodes.

65. Experiments with the Intensity Coil.—The experiments to be performed with the Intensity Coil are for the most part effects of the secondary current. They consist of luminous and calorific discharges, electro-chemical phenomena, inductive and physiological effects. These experiments, for the sake of perspicuity, will be grouped together, and some instructions given for performing the same. The phrase, "Attach to the Coil," will be frequently used in describing these experiments, and that will be taken to signify that the apparatus in question is to be connected with the binding screws that represents the two ends of the secondary Coil.

66. The Discharger.—In operating with an Intensity Coil, it will be found convenient, for some experiments, to employ a modification of the apparatus known as Henley's Discharger. This is represented in *Fig.* 66, and it consists of two glass or ebonite pillars rising up from a base board, and carrying

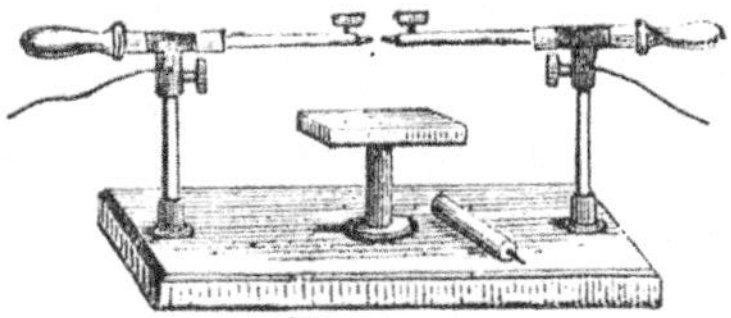

Fig. 66.

two binding screws and horizontal tubes, through which two brass rods with insulated handles slide. The ends of these two rods are drilled up a short distance, and are provided with side screws passing into the holes so made. The arrangement is made in order to admit of the attachment of small wires for the purpose of performing various experiments. The binding screws on the tops of the pillars are for connecting the Discharger with the Secondary Coil. A small ebonite table is fixed to the base board, and upon this table can be placed anything that is to be operated on. This table fits into a socket, and can be moved when required.

67. Primary Spark.—When the battery is connected with the coil, a small but intensely vivid spark is seen at the point of the platinum screw, where it comes into contact with the spring of the contact breaker (**51**). This spark appears as though it were a brilliant star, and when the battery and coil are large, it has the aspect of a flame. This, although apparently continuous, is in reality intermittent, as no spark exists while the platinum screw is in contact with the spring.

68. Secondary Spark.—"Attach two wires to the coil," and bring their ends near to each other. When the battery circuit is completed, a stream of sparks will pass between the wires, *Fig* 68. These sparks are manifestations of the dynamic condition of the secondary current (**14**).

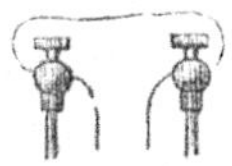

Fig 68.

69. Static Spark.—Remove the wire from the binding screws of the coil, and apply the knuckle or any other conductor to them. Sparks of a feeble brilliancy and of a comparatively loose appearance will past from *one* of them to the knuckle, such sparks being the manifestation of the static condition (**15**) of the current.

70. Elongated Dynamic Spark.—Attach two moderately thin silk-covered wires to the coil, and direct the *end* of one toward the *side* of the other, and when they are sufficiently near to each other long brush-like Dynamic sparks will pass between them. On examining the structure of the spark through a magnifier, it will be seen that the appearances of the two ends of the spark greatly differ. One end of the spark is the more brilliant, and of a red colour, the other end being of a feebly luminous violet colour, enveloping the entire end of the wire. These phenomena are more manifestly seen when the spark is passing through a vacuum (**116**).

71. Calorific Dynamic Spark.—Alter the position of the two wires, or electrodes (**64**), as they will in future be called, employed for the preceding experiment, and bring their *ends* opposite each other. Diminish the distance between the two wires, and the spark will become brighter and more intense, and assume a very firey red colour. This is the condition of spark suitable for ignitions, and all calorific effects.

72. Intermittent Character of the Spark.—Arrange the electrodes (**71**) so as to obtain sparks of medium length. These sparks follow so rapidly upon each other that the impression of one spark does not fade from the retina of the eye before another spark passes. That these are sparks and not a continuous flame, can be satisfactorily shown by the following experiment :—Direct the eyes to the stream of sparks with a somewhat vacant stare, and then, without moving the *head*, sharply turn the eyes either upwards or downwards. This movement will cause the image of the sparks to fall upon successive proportions of the retina, and a series of images, one below the other, will be seen, having the appearance of the rungs of a ladder.

73. Retaining Power of the Retina.—The impression made by a luminous object on the retina of the human eye lasts for about one-eight of a second; therefore, when the *eye* is rapidly moved, the impression is retained on the retina *after* the light from the spark ceases to fall upon that particular spot, and thus the series of luminous images result. When a continuous light (such as that from a taper passing through a small hole in a card) is treated in the same way, an elongated image is seen, but not a succession of separate images.

74. Divergence of the Spark towards a Flame.—Arrange the electrodes so as to obtain the spark of average length, and while the current is passing, bring the flame of a small taper near the stream of sparks. The character of the spark will be at once changed, its luminosity will be considerably diminished, and it will be attracted apparently by the flame, *Fig.* 74. The heated atmosphere surrounding the flame, and the particles of carbon in the flame, form a better conducting medium than the ordinary atmosphere ; and this is the probable explanation of the phenomenon.

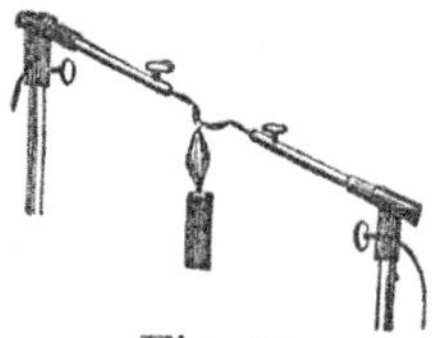

Fig. 74

75. Elongation of the Spark by a Flame.—Remove the taper from the *side* of the electrodes, *Fig.* 74, and place it with the flame fully below them. Under these circumstances it will be found that the electrodes can be separated much further apart than before. The spark becomes blue and loose, but three or four times longer than could otherwise be obtained. The rarefied condition of the air caused by the flame of the taper is the probable cause of this.

76. Action of the Flame of Bunsen's Burner.—Place a Bunsen's Burner on a sheet of ebonite or gutta-percha, and attach one of the electrodes to some part of the same. Light the burner, and approach the other electrode to the upper part of the flame, and the spark will pass down it to the metal tube from whence the flames issues. If a little common salt be thrown into the flame, its apparent conducting power is increased.

77. Change in Spark passing through a Flame. —Approximate the electrodes, *Fig.* 77, so as to obtain the calorific spark (**71**), and then bring the flame of a wax taper between them, so that the spark may pass over the top of the wick. The spark will assume an egg-like form of intense whiteness.

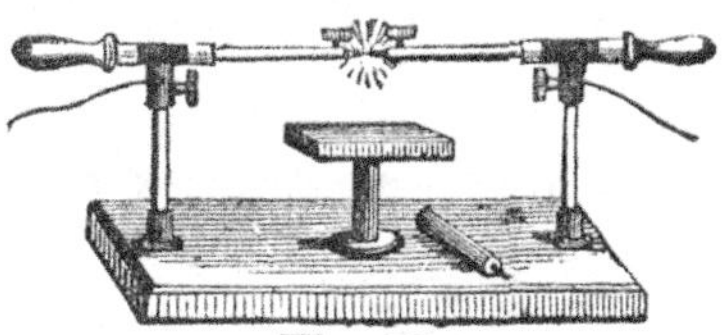

Fig. 77.

Separate the electrodes a little, and elevate the taper, so that the spark may pass through the base of the flame. It will be seen to travel round the wick, forming an intensely blue stripe.

In performing these experiments, the ends of the electrodes will be blackened by a deposit of carbon from the flame of the taper. When the taper is taken away, and the electrodes approximated towards each other so as to produce the calorific spark, this coating will be rendered incandescent, and the ends of the electrodes will glow like vivid white stars.

78. Modification of the Spark by Surface.—Attach a metal ball, about $\frac{3}{4}$ inch diameter, to one electrode, and approximate the end of the other electrode towards it, *Fig.* 78. Under these circumstances the spark will be divided into a number of branches radiating from the end of the electrode.

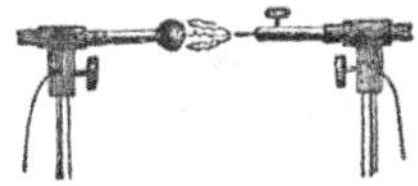

Fig. 78.

79. The Shield Experiment.—Remove the ball from the electrode, *Fig.* 78, and substitute a concave metal plate. A somewhate similar division of the spark will take place, but it will be accompanied by a loud report.

80. Apparent Bending of the Spark.—Arrange the electrodes so as to produce the elongated spark (**70**), and bring the edge of a gutta-percha or ebonite plate into contact with it, *Fig.* 80. The spark will be bent as though it were forced out of a straight line by the pressure of the plate against it.

Fig. 80.

81. Spark over Softened Gutta-Percha.—Approximate the two electrodes so as to produce more of a calorific spark (**71**), and then interpose a piece of thin gutta-percha. The spark will pass over the edge of the gutta-percha, which will become softened and partially transparent, and under these circumstances the zigzag form of the spark can be seen through the interposing sheet. In this softened state the gutta-percha offers less resistance to the passage of the spark over it, and consequently the edge of the gutta-percha can be moved further through the direct line of the spark.

82. Spark over Partial Conductor.—Moisten a plate of gutta-percha or ebonite, and then bring down the ends of the electrodes upon the wetted surface, *Fig.* 82. The spark under these circumstances will be greatly increased in length, and will assume a beautiful zigzag form, and be broken up into fragments or points of light.

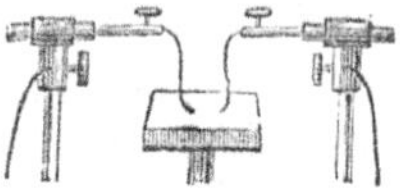

Fig 82.

83. Fiery Scrolls.—Lay a plate of tin-foil on a piece of ebonite or gutta-percha, and bring one of the electrodes into contact with the margin of the plate. Arrange the other electrode so that it may stand over the centre of the tin-foil, leaving a distance between something more than that neccessary for the calorific spark (**71**). The stream of sparks appears as though it was separated into a number of diverging rays, which, impinging on the plate, become formed into a snake-like shape.

84. Sheet of Fire.—Approximate the electrodes so as to obtain the calorific spark (**71**), and while the stream of electricity is passing, blow upon it through a glass taper tube. The tube may be attached to a small pair of bellows, or it may be blown through from the mouth.

The spark itself appears unaffected, and proceeds without any disturbance from the blast of air passing over it. But a large sheet of fiery-coloured flame is driven out from the opposite side, apparently as though the air passing through the stream of sparks became ignited.

85. Spark over Finely-divided Metal.—Place some copper filings and steel filings in a mortar and rub them down very fine. Sift the combined dust lightly over a piece of ebonite or gutta-percha. Bring the two electrodes in contact with the metal filings, and the spark will dart in all directions, deflagrating in its passage in some of the particles of metal, and acquiring thereby an artificial colour.

86. Spark over Water.—Place a large drop of water over the surface of a glass or smooth ebonite plate, and bring the electrodes down towards the opposite margins of the water, *Fig.* 86. By this arrangement the spark will pass through a much larger space than it would if the water were removed. The spark passes over or round the edge of the water, and not through it.

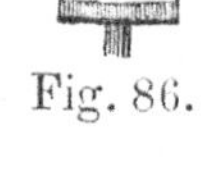

Fig. 86.

87. Spark obtained from Water.—Place one of the electrodes into an ordinary drinking glass filled with water. On bringing the end of the other electrode towards the surface of the water, vivid sparks will pass between it and the water.

88. Action of Oil on the Colour of the Spark.—Dip

the ends of the electrodes in oil, and then approximate them so as to obtain the calorific spark (**71**); the colour of the spark will be intensely green.

89. Spark not Extinguishable by Water.—To perform this experiment it is necessary to prepare two small glass tubes about two inches long, with one end of each bent at right angles. Through these tubes platinum wires are to be passed, so that they just come through the bent end of the tube. The glass at this end is to be carefully melted round the platinum, and when cold any projecting wire is to be cut off, so that a surface of platinum equal only to the area of the section of the wire remains exposed. These tubes are to be fitted through a disc of cork, in such a way that the platinum ends may be opposite to and all but touching each other, *Fig.* 89.

The disc of cork carrying the tubes is now to be laid on to the top of a glass vessel filled with distilled water, and the electrodes from the coil laid upon the platinum wires projecting from the upper ends of the two glass tubes. As soon as the coil is in action, small vividly white sparks will pass through the water.

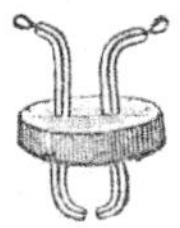

Fig. 89.

90. Influence of the Electric Current upon Oil.—Place upon the plate, *Fig.* 90, a globule of oil, of larger or smaller dimension according to the power of the coil employed. Bring down one of the electrodes near to the margin of the oil, and a slight commotion will be observed in it. It appears as though that side of the oil were in a state of effervescence. Place the other electrode on the opposite side, and likewise near to the margin of the oil. The effervescence will be remarkably intensified, and will manifest itself in apparent flashes, radiating from one electrode towards the other electrode, and pervading more than half the distance between them. These flashes are not electrical, but are only reflections of the light from the room, which, falling upon the oil, shows the perturbation of its surface.

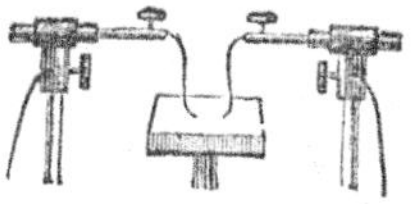

Fig. 90.

The margin of the oil opposite one of the electrodes will begin to thrust itself out, apparently attracted by it, and this will go on until the electrode is reached and enveloped. The side of the globule of oil towards the electrode now seems disintegrated, and appears to be driven towards the other electrode, a thin film only remaining, which appears as though it were dried by heat. This is probably the explanation: the resistance offered by the oil to the passage of the current resolving some of the electricity into heat.

91. Apparent Impelling Power of the Current.—Place a small quantity of some light non-conducting powder on a plate of opal glass, and make a slight mound of it by gently tapping the under side of the glass. Let the base of the mound be more in diameter than the space through which the spark of the coil to be employed will pass. Place the ends of the two electrodes on the opposite edges of the powder, and barely in contact with it.

As soon as the coil is in action, the particles of powder are seen to be in motion. One electrode seems as though it blew away the particles near to it, and a division begins to appear in the edges, midway between the two electrodes. This rent increases until it reaches across the mass of powder, dividing it into two parts. The portion of one side continues to diminish by the particles being transferred across the channel between the two portions to the other side.

But the particles do not stop here; they are spread far and wide over the glass round the electrode, and arranged smoothly and regularly as a snow drift. It is a noticeable thing that, although the particles seem thus blown away, yet a small quantity gathers round the electrode, completely enveloping it some distance up from the end, while, at the same time, although the particles are apparently blown towards the other electrode, yet the glass round its end is completely clear, as though it repelled all the particles immediately contiguous to it. Crystallized gallic acid is a very good powder to use for this purpose, as it consists of minute needle-shaped crystals that are easily observed through a magnifier, and, consequently, the movement of the particles is clearly seen.

92. Dr. Wright's Cohesion Experiment.—For this experiment, a clean plate of mica or glass is to be laid upon a metal plate, which should be, for the better display of the result, dead black. One electrode is to be attached to the plate, and the other placed vertically in the centre of the mica or glass, and in contact with it. A drop or two of liquid is to be placed on the glass round the end of this electrode. As soon as the coil is in action, a number of small points form at the margin of the drop of liquid, which push forward and become trunks from whence other tiny streams start. After continuing the action for some time, the glass plate is covered with the most beautiful arborescent figures.

It is not of consequence which of the electrodes be in contact with the liquid, the only difference being a change of pattern. Different figures are produced by water, acid, and saline solutions, and the same solution acts differently when placed upon mica or upon glass. The figures produced seem to be regulated by the cohesion between the fluid and the surface upon which it rests, and also by the cohesion of the particles of the fluid itself.

93. Disruptive Action of the Spark on Paper.—Lay a plate of tin-foil or other metal on a suitable support, and attach one of the electrodes to it. Place on this a piece of writing paper, and bring the end of the other electrode near it. As soon as the spark passes the paper is punctured, and by removing the electrode, and approaching it again to other parts of the paper, a series of holes can be made, and any required pattern can thus be pierced. If card can be used instead of paper, an appreciable space of time must elapse before the electrode is removed, in order thoroughly to perforate the card.

94. Peculiar Character of these Perforations.—It is worthy of note that these holes are cleanly pierced, without any burr on either side, the margin of the holes being blackened as though the paper were charred. In this respect they differ from the perforations made in paper or card by the charge of an ordinary Leyden jar. In this latter case there is a remarkable burr round the holes, as though they were punctured with an instrument that only thrust aside the fibres of the paper, and did not remove the portion necessary to leave the hole.

95. Decomposition of Water.—This is shown by employing a cylindrical glass vessel, through the sides of which two platinum wires are passing, which should proceed some distance towards the centre of the cylinder, and then turn up at right angles, *Fig.* 95. The cylinder thus prepared should be partly filled with water, well acidulated with sulphuric acid or vinegar, and it will then be ready for use. The two electrodes of the coil are to be attached to the ends of the platinum wires outside the cylinder, and, as soon as the coil is in action, small bubbles of gas will be seen to form on the wires in the water. The bubbles are oxygen gas and hydrogen gas, and are the components of water.

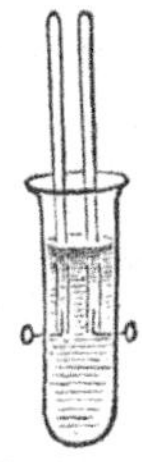

Fig. 95.

If a cork carrying two tubes closed at one end be fitted into the mouth of the cylinder, so that the open ends of the tubes come over the platinum wire, and the tubes be filled with the acidulated water by inverting the whole apparatus, the hydrogen gas will ascend into one tube and the oxygen gas into the other.

96. Decomposition of a Neutral Salt.—Dissolve starch in hot water, and dip into it strips of stout white blotting-paper, and hang them up to dry. When dry, moisten the surface of one of the papers with a weak solution of iodide of potassium, and while wet place the paper upon a plate of glass or ebonite. One of the electrodes of the coil is to be attached to the margin of the paper, and the other electrode, which in this case must be armed with platinum, is to be brought in contact with its surface. As soon as the coil is set in action, a purple spot is seen at the end of this electrode, resulting from the decomposition of the iodide of potassium and the liberation of the iodine, which unites with the starch in the paper, forming iodide of starch, the colour of which is purple.

By employing an electrode with a smooth end, letters or devices may thus be produced, the electrode apparently writing or tracing a purple line (**44**).

97. Scintillation of Iron Wire.—Affix a short length of fine iron wire to each electrode, and bring the ends near together so as to obtain the calorific spark (**71**). The end of the wire attached to one electrode will become red-hot and ignite, while that at the other electrode will remain cool, *Fig.* 97. If the wires be surrounded by a glass tube about $\frac{3}{4}$ inch diameter, the cooling action of the atmosphere will be lessened, and the deflagration intensified.

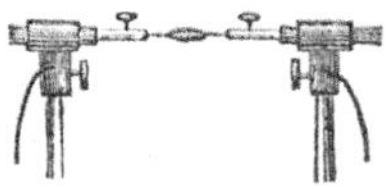

Fig. 97.

98. Deflagration of Leaf Metal.—Place a leaf of silver or gold or Dutch metal upon a sheet of paper, supported on ebonite or gutta-percha. Let one of the electrodes be in contact with one edge of the leaf, and approach the other electrode to the opposite edge. The metal deflagrates and rapidly disappears, colouring the spark, and leaving an indelible stain upon the paper.

99. Ignition of Phosphorus.—Place a piece of phosphorus of the size of a small pea upon a glass plate, and allow the spark to pass through it. The phosphorus will be ignited, and burn with the usual dense white fumes. The phosphorus should be on a piece of glass or porcelain, sufficiently large to prevent portions of the ignited phosphorus falling on to the table.

100. Inflammation of Ether.—Place a few drops of sulphuric ether upon a tuft of cotton wool, and by the aid of a pair of forceps hold the tuft of wool between the ends of the electrodes while the sparks are passing. The ether will be instantly ignited. The spark should be of a medium length, approaching to the character of the calorific spark (**71**).

101. Inflammation of Hydrogen Gas.—Arrange the electrodes so that sparks of medium length will pass between them. Direct a stream of the gas from an ordinary gas-burner upon the sparks, and it will instantly burst into a flame.

102. Explosion of Gas and Common Air.—For this purpose, a brass tube, shaped like a pistol barrel, with a wooden stock, *Fig.* 102, is the most convenient apparatus. An ivory or ebonite plug must pass through one side of the barrel, carrying a wire that should come near to the opposite side of the barrel.

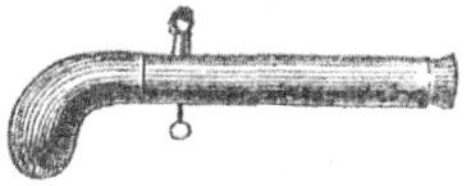

Fig. 102.

The barrel is to be charged by holding its mouth downwards over an ordinary gas burner; the mouth is then to be closed with a cork. One electrode is to be attached to an eye in the barrel itself, and the other to the wire passing through the plug; the spark will then pass through the mixed gas and air and ignite them, and blow out the cork. In performing this experiment some care should be exercised, otherwise the barrel will be *entirely* filled with hydrogen gas. If this be the case it will not inflame.

103. Ignition of Gun Cotton.—Hold a tuft of gun cotton by means of a pair of forceps in the stream of sparks, and it will instantly ignite and disappear with a feeble report.

104. Combustion of Lycopodium.—Scatter some lycopodium over a small tuft of cotton, and place some of the filaments in the stream of sparks; the lycopodium will be set on fire.

105. Re-lighting a Taper.—Ignite a taper and then blow it out, and when it has ceased to smoke bring the wick between the ends of the electrodes while the sparks are passing, and it will be re-lighted.

106. Explosion of Gunpowder.—For this experiment the gunpowder should be finely pulverised and sprinkled on cotton wool, as was directed for the lycopodium (**104**). As soon as some of the filaments of the cotton carrying the gunpowder are brought into the stream of sparks the whole is ignited.

107. Galvanic Cartridge.—Twist together two short lengths of gutta-percha covered wire, leaving a portion of the wire at each end uncased. Bend the wires at one end a little bow form, so that the end m stand opposed to each other and about $\frac{1}{8}$ inch apart. Cut off a short length of gutta-percha tube about $\frac{1}{2}$ inch in diameter and 1 inch long. Soften one end of the tube and place it over the bowed end of the twisted wires, and press down the sides of the tube that it may come in contact with the gutta-percha coating of the wires, and unite with the same. The cylinder thus formed, *Fig.* 107, is to be filled with fine gunpowder, and the end closed with a disc of gutta-percha. When this is neatly done the gunpowder will be enclosed in a water-tight chamber in contact with the ends of the wires. This may now be placed in a vessel of water, and as soon as the outside ends of the twisted wires are attached to the electrodes the cartridge will be exploded.

Fig. 107.

108. Statham's Fuse.—To form this fuse it is only necessary to take a length of copper wire, covered with vulcanised india-rubber, and fold it back upon itself and twist it together. The wire, near where it is bent, will require to have a portion of its covering cut away, so as to enable a part of the wire to be removed, leaving a break in it of about $\frac{1}{4}$ inch in length. This must be carefully done, so as not to damage the inside layer of the remaining covering. This portion of the covering should now be wrapped in a tuft of gun cotton and then protected by a suitable case. As soon as the two ends of the twisted wires are attached to a battery the gun cotton is exploded. The explanation is this: The contact of the vulcanised covering with the copper wire forms on the inside of the covering a coating of sulphide of copper; this is an imperfect conductor, and the current, intercepted in its flow by the break of the wire, proceeds over the coating of sulphide, which becomes ignited, and so inflames the gun cotton.

109. Abel's Fuse.—This consists of a tube or chamber having two wires passing into it. The ends of these wires must be further apart than the distance capable of being traversed by the spark of the coil it is to be used with. The chamber is charged with a mixture of chlorate of potassa, subphosphide of copper and subsulphide of copper. These ingredients require to be mixed in proportions that will only partially conduct the current, in consequence of which, when the fuse is connected with the coil, heat is developed and the fuse ignited.

110. Charging a Leyden Jar—Insulate a Leyden Jar by placing it upon a plate of ebonite or other non-conducting substance, and attach one of the electrodes to the ball of the jar. The other electrode is to be placed some little *distance from* the outside coating of the jar, and as soon as the coil is in action the jar begins to charge, and it can be discharged by an ordinary discharging rod.

This experiment is more interesting if, instead of employing a Leyden Jar with continuous tin-foil coatings, one covered with diamond-shaped spots be used. The diamonds on the *outside* of the jar should be pierced through the centre, so as to allow the points of the diamonds on the inside of the jar to be readily seen.

111. Detonating Plane.—This piece of apparatus consists of a plate of glass about 1 foot square, having a coating of tin-foil on each side, with about 1 inch all round left uncoated. The plate should be supported on its edge on a sheet of ebonite or gutta-percha and one of the electrodes brought into contact with the tin-foil coating of one of its sides. The other electrode is to be brought near the opposite coating, leaving about $\frac{1}{4}$ inch space between them, *Fig.* 111.

As soon as the coil is in action, sparks will be seen to pass between the electrode and the tin-foil, and in a few seconds a vivid flash will dart from one coating over the uncoated edge to the other coating, accompanied with a loud report. This report will be repeated at intervals as long as the coil is in action. The plate is, in fact, a coated Leyden surface, and the sparks over the edge are caused by the spontaneous discharge of the overcharged surfaces of the plane.

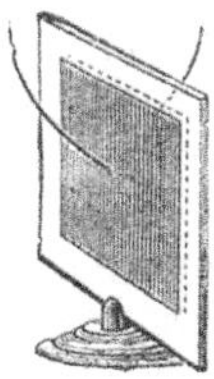

Fig. 111

VACUUM EXPERIMENTS.

Fig. 112.

112. Spark in a Rarefied Medium.—The foregoing experiments are interesting and important in a scientific point of view, and many of them are extremely beautiful; but it is when the electric discharge passes through transparent vessels partially exhausted that the incomparable splendour of these electrical phenomena is seen.

For such experiments the vessel must be of glass, and metallic wires must be inserted air-tight at the two ends, or at some suitable distance from each other, a portion of the wires remaining outside the glass vessel. The vessels may be large in diameter, or may be diminished to the size of the tubes. They may be straight, or bent and curved in any conceivable device, *Fig.* 112. They may be of ordinary glass, or of glass having certain metallic oxides in its composition. The vessels or tubes may be filled with the atmosphere in a highly rarefied state, or they may be filled with other gases, inorganic or organic, in a similar condition. If there be a perfect vacuum in the vessel the spark will not pass at all, nor will it pass if the vessel be full of air of the ordinary density; a highly attenuated medium of some kind is essential.

The following experiment demonstrates the necessity of an atmosphere within any vessel that is to be made a pathway for the electric current:

For this purpose a small tube only is required, one about 3 inches long and $\frac{1}{2}$ inch diameter will do very well. The metallic wires should pass into the middle of the tube, so that their ends may almost touch each other. The tube so prepared is to be perfectly exhausted, and on connecting it with a coil, there will not be the faintest trace of a spark between the ends of the wires within the tube, although these may not be more than ·01 inch apart.

In attaching such a tube to a coil it is usual to employ the discharger (**66**) for the purpose. In that case the rods of the discharger should be placed obliquely to each other across the base board, that the ends may be approximated so as to obtain a calorific spark (**71**). The exhausted tube is now to be suspended to the rods of the discharger by means of fine uncoated wires passing through the projecting wires on the ends of the tube. It will be seen that this arrangement establishes *two* modes of completing the circuit: one by the spark leaping the interval between the ends of the rods of the discharger, and the other by passing between the ends of the platinum wires within the tube; the space in one case being perhaps 3 inches, and the other case 01 inch; the largest space being, moreover, a space in air, and the smallest being a space in vacuo.

On attaching the rods of the discharger to the coil the spark will pass in a stream between the ends of the rod; but not the least trace of a spark can be seen inside the tube, thus proving that the electric discharge will not take place in a vacuum.

113. Luminous Electric Globe.—Prepare a glass globe 5 or 6 inches in diameter, having a stopcock at one end and a cap at the other. Let a wire with a rounded end project from inside of the cap at the top of the globe, and also from the end of the stopcock, likewise within the globe. When completed, the apparatus will be a glass vessel with a cap at each end, and with two rounded wires projecting opposite to each other, a short distance within the globe, *Fig.* 113.

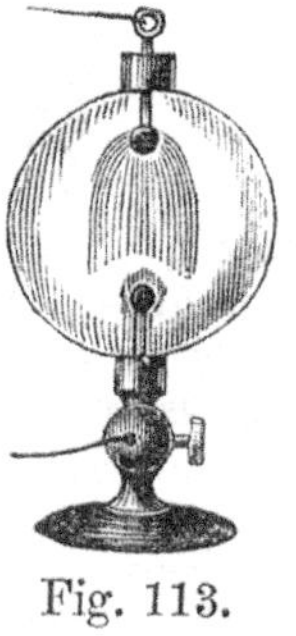

Fig. 113.

Let the globe be partially exhausted, and then connect the two electrodes of the coil with the two caps of the globe, and put the coil into action. Instead of a spark it may be $\frac{3}{4}$ inch or 1 inch long, such as could be obtained between the electrodes of the coil, a splendid stream of light is seen reaching across the inside of the globe from one wire to the other.

114. Stratification of the Spark.—Detach the globe from the coil, and exhaust it thoroughly, and then introduce a small portion of the vapour of alcohol, naphtha, or any similar hydrocarbon. Connect the globe again with the coil, and it will be found that a great change has taken place in the character of the light. Instead of appearing now as a stream pouring from wire to wire, it will be divided into zones or bolts, *Fig.* 114.

Frequently these zones appear as though they were slowly revolving on their own axis, having dark striæ over their edges. Such relations are apparent only, for, from some experiments of M. Quel, there is reason to believe that such a movement is only an optical illusion.

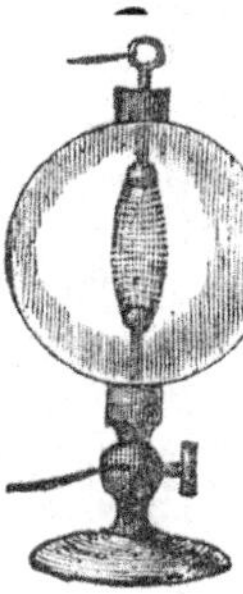

Fig. 114.

115. Attraction of the Spark by a Detached Conductor.—Arrange the luminous globe as in **113** or **114**, and while the stream of light is passing through the globe hold the metal wire or other conductor to the outside of the globe, *Fig.* 115. The current will be diverted from its path and flow to the side of the globe where the conductor is.

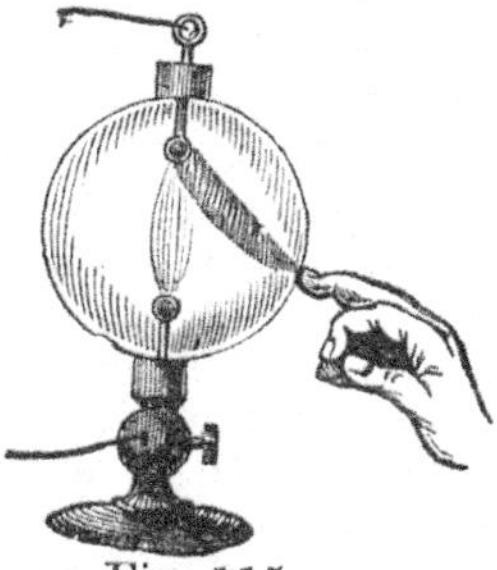

Fig. 115.

116. Different Appearances at the two Wires.—It has been before pointed out (**70**) that the two ends of the electric spark passing between the electrodes of the coil greatly differed in appearance. This is seen in a much more palpable and beautiful manner in the exhausted vessels. The wires projecting into these vessels are, in fact, prolongations of the electrodes, and the appearance at each is most striking.

At the electrode, which may be in effect designated as the anode, the manifestation is that of a vivid star, usually of a fiery red colour, just at the end of the wire. At the cathode it is usually of a violet colour, soft, and diffused in appearance, and enveloping the sides of the wire as well as its end. Frequently it looks not much unlike a violet bulrush.

117. Change in the Colour of the Spark.—Exhaust the globe, and then open the stopcock so as to admit a very small portion of the atmospheric air, and connect it with the coil. The prevailing colour of the stream of light now seen will be mauve. Again exhaust the globe, and admit a small portion of hydrogen gas. The colour under these conditions will be mostly white. Once more exhaust the globe, and admit a like quantity of nitrogen, and the colours given will be rose, verging to deep orange. If the globe be again deprived of its atmosphere, and a small portion of carbonic acid be admitted, the colours obtained will be pale green.

118. Fluorescence—This term, fluorescence, is applied by Mr. Stokes to the property that some bodies have of altering the refrangibility of certain rays of light that fall upon them. This causes rays of light that are produced by undulations too rapid to be appreciated by our visual organs to become visible: and also, by a like process, alters the colours of other rays.

119. Fluorescent Plate.—Examine a plate of glass, having oxide uranium in its composition, by any ordinary light. It will appear of a feeble canary colour. Place it near an electric globe (**113**), or tubes (**123**), while the light is streaming through, and the uranium glass will be seen of a brilliant green colour.

120. Fluorescent Writing.—Dissolve a portion of sulphate of quinine in water, and write with the solution on a piece of paper. When dry, and viewed in ordinary daylight, the writing is all but invisible; but when the paper is illuminated by the light from the electric globe (**113**) or tube (**123**), the paper will appear as if written with beautiful blue ink.

121. Phosphorescence.—The word phosphorescence is used to express the properties that some substances, organic and inorganic, have of emitting light when there is no combustion. It is also employed to signify the curious power that some bodies have, when they have been exposed to light, of retaining the capability of exciting luminous undulations, and thus becoming visible in a dark room.

122. Phosphorescent Tube.—Fill a glass tube with dry sulphide of calcium, or sulphide of strontium, and cork it up. View such a tube by the aid of light from the electric globe (**113**) or tubes (**123**), and no particular appearance will be manifested. Remove the tube to a dark room, and the powder will then be seen luminous with phosphorescent light (**132**).

123. Vacuum Tubes.—M. Geissler, of Bonn, first constructed the tubes designated "Vacuum Tubes," and they have since been largely manufactured in this country, being better known in England as "Gassiot" tubes. They consist of closed tubes, with two platinum wires passing through their sides or ends, carefully melted in, and they are filled with the required vapour, and then hermetically closed. These tubes admit of all but endless variety of figures, and therefore the description of a few special forms will be sufficient.

When a Vacuum Tube is to be attached to a coil, one end of each electrode should be ben into the form of a hook, so as to unite more advantageously with the two projecting platinum wires in the tubes.

124. Gassiot's Cascade.—This is one of the most beautiful experiments that can be performed. The usual form of the apparatus is that of a cylinder of glass, slightly enlarged at each end, bulb form, *Fig.* 124. The upper bulb is divided from the rest of the cylinder by a glass partition, in the centre of which is a tube reaching down towards the lower bulb. This, like the upper, is separated by a glass partition, having one or more holes through it. On this lower partition is fixed a glass goblet, so arranged that the tube from the upper bulb comes down inside the goblet nearly to its bottom. The platinum wires for the electrodes are fixed one in the upper and the other in the lower bulb.

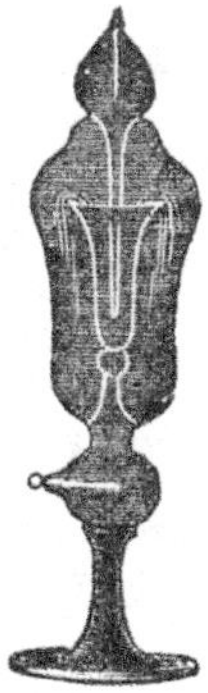

Fig. 124.

When such a vacuum tube as this is connected with a coil, the spark enters the upper globe and diffuses itself there. Then it passes down the centre tube and merges from its end on to the bottom of the goblet, rises from here up the inside of the goblet, pours over its edge, streams down the outside through the hole in the bottom partition into the lower bulb, and finally escapes by the platinum inserted there. The goblets are usually made of uranium glass (**119**), so that they glow with a beautiful green light, as though they were self-luminous.

125. Compound Vacuum Tubes.—This phrase is intended to signify a vacuum tube, of any desired form or device, enveloped in a glass chamber. This chamber has no connection with the interior of the vacuum tube, its purport being to form a hollow case or covering round the outside of the tube. The space between the two is filled with an aqueous solution of sulphate of quinine, or an alcoholic solution of stramonium, or a solution of æsculine. These solutions being fluorescent greatly add to the beauty of the device (**118**), communicating a coloured margin (usually a bright blue) to the stream of light that passes through the tube.

126 Compound Volute Tube.—This tube is represented at *Fig.* 126, which may be taken as a type of the class of globe tubes. The general arrangements of these tubes may be thus described:—The central portion is a large globe, containing within it a convoluted tube. This tube is sometimes, as in the illustration, a simple volute; other arrangements have more complicated figures, such as crowns, crosses, stars, &c. The central globe holds the tube, but is not connected with it;

it forms, in fact, a chamber that is filled with one or other of the fluorescent solutions described (**125**). The two ends of the convoluted tube enter into two smaller spheres that contain the platinum wires. When such a tube as this is attached to the coil, the appearance is beautiful: the upper and lower globes are filled with a soft, diffused light of rose or violet colour. The convoluted portion, occupying the central globe, shines out with vivid brilliancy, of uniform or varied colours, according as the glass of which it is composed is of one or mingled kinds. The fluorescent solution in the central globe gives a splendid fringe to the convolutions, and causes the whole to glow with coloured light.

Fig. 126.

127. Compound Spiral Tube.—This tube is shown at *Fig.* 127, and is only a modification of the form of that previously described (**126**). There is the interior tube, of a zigzag or spiral form, having at each end a bulb, with the platinum wire, and surrounded with a containing cylinder for holding the fluorescent solution (**125**).

Fig. 127.

128. Double Branch Compound Tube.—The illustration given at *Fig.* 128 represents this class of tube. Each arm has the central convoluted tube, one end of which is attached to the upper globe, and the other end to the U-shaped tube that joins them together. Each branch is surrounded with the enveloping chamber for the fluorescent solution (**125**), and the convoluted tubes are made of similar devices in each branch, and also with dissimilar devices. When the forms of the convoluted tubes in the two branches are alike, different varieties of glass are sometimes used, and sometimes different fluorescent solutions are employed in the two branches.

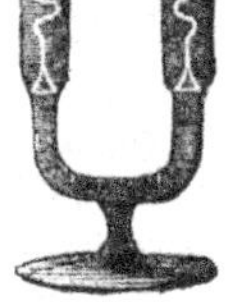

Fig. 128.

It will be obvious that the vacuum tubes can thus be varied almost indefinitely, every change in form or arrangement introducing a fresh variety of the most lovely combination of colours that can be well imagined.

129. Horizontal Compound Spiral Tube.—Some devices show more advantageously when the tubes are horizontal, and these may be adverted to. At *Fig.* 129 one such tube is

Fig. 129.

represented with a spiral convoluted tube, surrounded by a single or double chamber for holding the fluorescent liquid (**125**).

130. Horizontal Compound Bulb Tube.—This is

Fig. 130.

shown at *Fig.* 130, and it consists of a series of short tubes, opening into a succession of bulbs, and surrounded by the enveloping chamber of the fluorescent fluid (**125**). This form of tube is varied by using different kinds of glass, as well as by the number and size of the bulbs.

131. Motto Tubes.—Specimens of these tubes are given

Fig. 131.

at *Figs.* 131 and 131*, The convoluted tube is bent into the shape of letters, forming in this case the motto, "God save the Queen." There is the enveloping chamber for the fluorescent fluid (**135**), and when such a tube is attached to a coil the motto appears as though written in flaming letters.

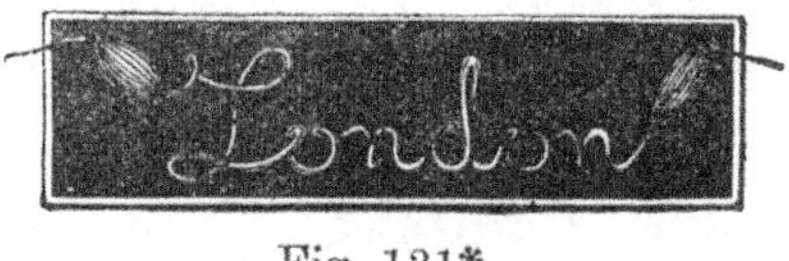

Fig. 131*

The second illustration, *Fig.* 131*, is without the enveloping chamber, and the bulbs are blackened, so that the device only is seen.

132. Phosphorescent Vacuum Tube. — This is usually a cylindrical tube, with the platinum wire at each end, and containing dry sulphide of calcium. When attached to the coil, the inside of the tube is filled with a splendid stream of light varying in colour with the nature of the gas contained in it (**117**), the sulphide of calcium in the tube appearing like a reflecting surface, adding to the brilliancy of the light. On detaching the battery from the coil, or otherwise interrupting the passage of the electricity, the stream of light in the tube is instantly extinguished, but the sulphide of calcium continues to glow with phosphorescent light (**122**).

133. Seven Bulb Tube.—This is a beautiful kind of tube, and is usually made in the form of a triangle, with three bulbs on one side and one bulb on each of the other sides, the sixth and seventh bulbs carrying the platinum wires.

The appearance of this tube is very fine. The bulbs are filled with soft diffused light streaming round, and the connecting tubes look like intensely gleaming bands of fire.

134. Six Bulb Tube.—The usual form of this beautiful tube is that of six bulbs in a row with U-shaped glass tubes between them. By this arrangement the electricity is caused to enter the first bulb by the platinum wire, and pass through the U-shaped tube to the upper part of the inside of the next bulb. The current then streams round the bulb in order to find an exit at its lower side, when it enters the U-tube, and so on to the end of the series. The light in the bulbs is soft and diffused, while that in the U-tubes is vivid and brilliant, and usually manifests most strikingly the appearance of stratification (**114**).

135. Double Bell Tube.—This elegant tube, in its exterior form, consists of three bulbs united together by large cylindrical junction tubes. The first and last bulbs are separated from the other parts of the tube by partitions of glass, from which partitions proceed small short tubes reaching nearly to the central bulb. These tubes terminate in contracted bell-shaped ends. When such a tube as this is attached to a coil, the first and last bulbs are beautifully illuminated with soft violet light, and at the same time the interior tubes glow with vivid light, which streams out of the mouth of the one bell, and after proceeding in a diffused form through the central bulb, enters the open mouth of the opposite bell.

136. Combined Bulb and Spiral Tube.—This may be described as two bulbs united together by a convoluted tube. This tube, for a short distance from each bulb, consists of a series of small bulbs with straight intervening tubes, the rest being twisted into a beautiful helix. An enveloping tube is usually provided for holding the fluorescent fluid (**125**); and the convoluted tube can be made of different qualities of glass, so as to give increased variety to the colours of the light.

137. Nine Bulb Tube.—This tube is made in the form of a circle, about 6 or 7 inches in diameter, and presents the appearance of a large central bulb surrounded by a ring having four large and four small bulbs arranged symmetrically upon it. This is one continuous channel, and, when illuminated, presents a ring of glowing bulbs hung upon a brilliant cord.

138. Pagoda Tube.—This tube consists of the nine bulb

tube (**137**), supported upon a spiral cone. The effect of such a tube as this is really magnificent, showing one mass of glowing variegated light.

139. Triple Bar Tube.—This, like other of the complicated tubes, is difficult to describe. It consists of a centre cylinder, with a series of small bulbs at each end. The ends of the cylinder are contracted and united to smaller tubes, which are bent so as to form a tube on each side of the central cylinder. These side tubes have a series of bulbs arranged on them, and they each eventually terminate in a large bulb. When illuminated, this complex tube has three lines of light parallel to each other, with the varying effects of fourteen bulbs of different sizes interspersed throughout its length. The effect of these combinations of figures is most beautiful and surprising.

140. Gassiot Star.—This is an apparatus for multiplying the effects of a single tube, and it accomplishes this by causing the tube to pass rapidly before the eye. The vacuum tube is attached to a wheel, *Fig.* 140, so that it may be rotated, the axes of the wheel being provided with proper arrangements for keeping the tube in contact with the coil. When the wheel is rapidly rotated the impression is retained on the retina of the eye, and, instead of one, a number of tubes are seen, forming the radii of a star.

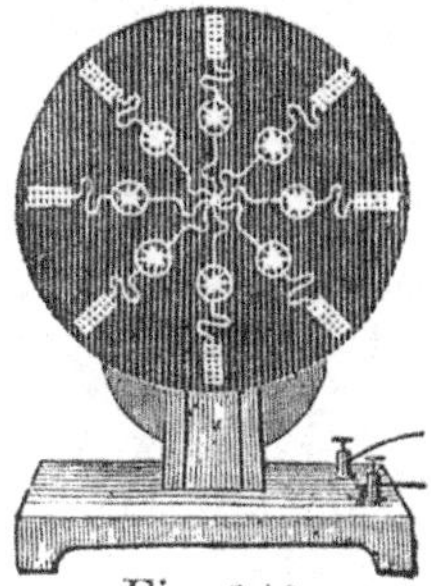

Fig. 140.

The wheel can be revolved by a handle and band, or it can be attached to a magnetic rotation machine, in which latter case it is self-acting.

141. Multiplying Planes.—This is an arrangement for increasing the effect of the vacuum tubes by multiplying their images by reflection. The apparatus consists of two slips of silvered glass of a length equal to that of the longest tube that is to be used with it; the width of the glass plates should be about one-third of their length. The plates are set up vertically with their faces towards each other, standing at an angle of 60 deg. one with the other. They are to be fixed in this position, and are to be provided with suitable connections, so that the tube may be held upright between them, and attached to the coil. When the tube is illuminated, the image is reflected from both surfaces, and it is as though six tubes were being looked at instead of one only.

142. Injury to Vacuum Tubes.—It will be obvious from these descriptions of vacuum tubes that they are necessarily fragile articles, and require to be handled carefully, so as to guard them against breakage. But over and above such danger as this, they are liable to damage from the action of the electrical current itself.

If the power of the coil be greatly disproportioned to the length of the tube, it not unfrequently happens that the platinum wire becomes melted off inside the tube, and the tube spoiled.

Even though the tube and the coil be duly proportioned to each other, the end of the tube attached to the cathode will often become heated, and therefore the direction of the current ought to be reversed, or the tube should not be used for any length of time without periods of cessation, so as to admit of the end becoming cool.

143, Union of Vacuum Tubes.—Tubes with comparatively short circuits can be united together, and fresh varieties can by these means be obtained. The only limit to such a union of tubes is the power of the coil. Each additional tube obviously lengthens the path through which the current has to travel.

144. Back-ground for Vacuum Tubes.—Where it is practicable, it is best to have a black velvet screen behind the tubes, and if they can be also screened from top and side light the beauty of the exhibition will be greatly enhanced.

145. Physiological Effects of the Intensity Coil.—These effects are obtainable from the primary as well as the secondary circuit, and they are most remarkable. When the force of the current is modified by reducing the battery power or interposing a partially non-conducting medium between the conductors and the hands, or other part of the body to which the current is to be applied, the sensation experienced is that of a rapidly repeated throbbing. By increasing the force of the current the gentle throbbing becomes more decided, and as the power is augmented assumes a character of rigour. The muscles become contracted, and from a pleasant, almost soothing sensation, one of extreme pain and agony is reached, which soon becomes intolerable.

146. Shock from Primary Circuit.—To obtain the primary shock, a separate or loop circuit must be established. This is effected by attaching a wire to the pillar carrying the platinum screw of the contact breaker (**58**), and fixing a similar wire to the block that carries the spring and clapper. To the end of these two wires metal cylinders or plates are to be fixed, and these are to be grasped in two hands, or laid upon that part of the body which is to be acted upon.

It will be seen from this that the primary circuit has now two channels, one through the contact breaker when the platinum screw is in contact with the spring, and at the same time a second channel by means of the wires just described and the person to be operated upon. But the conducting powers of these two greatly differ, and the consequence is, that so long as the platinum screw remains in contact with the spring, a perfect metallic circuit is thereby provided, and no current passes through what may be called the loop circuit. But as soon as the metallic circuit is interrupted by the severance of the spring of the platinum screw the loop channel is brought into operation, and a shock is felt.

This shock is experienced every time the contract breaker vibrates, and the rapidity with which this takes place causes the shocks to follow each other in such rapid succession that an almost continuous effect is experienced.

The reason why a shock is obtained when the battery current is passed through a coil has been before explained (**40**), and need not be repeated here.

147. Uni-direction Shock.—The effects just described are uni-direction shocks ; the induced current flowing in the same direction as the battery current. Such a condition is of importance when the coil currents are to be employed for many physiological operations. The experiments previously detailed (**44, 96**) showed that many results are only obtainable at one of the two electrodes, and consequently any attempt to obtain physiological results analogous to the effects there described will fail unless the coil be so constructed as to give uni-direction currents.

148. Shock from Secondary Current.—To obtain these shocks from an Intensity Coil it is only needed to fix the two wires with the metal cylinders to the electrodes of the Secondary Coil and grasp the cylinders in the hands. The shock thus obtained from even the smallest Intensity Coil will be too strong to bear at all pleasantly, and therefore a modifying arrangement, such as the water regulator, must be resorted to. Ordinary coils are so constructed as to admit of such methods of regulation being part of their formation. Intensity coils, however, cannot be so made, and therefore a separate instrument is required.

149. Water Regulator.—This is a glass cylinder about half-inch internal diameter, and 5 to 12 inches long, fixed into a metal cap upon a foot. The upper end of the tube is also closed with a metal cap, through which a metal rod slides, *Fig.* 149. The length of the sliding rod should be something more than that of the tube, so that when occasion requires, the rod may be brought into contact with the inside of the cap forming the base of the tube. The glass tube is to be filled with ordinary soft water, and then it is ready for use. The binding screw in connection with the bottom cap is to be attached to one of the electrodes of the secondary coil, and the wire from one of the metal cylinders

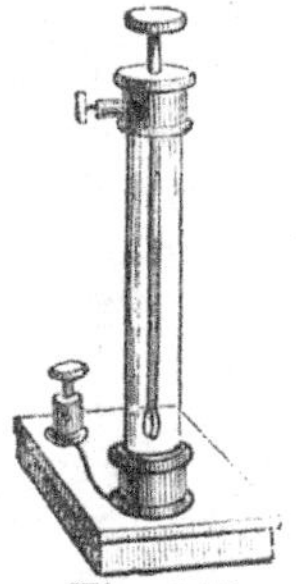

Fig. 149.

is to be attached to the sliding rod by means of the binding screw on the top cap.

It will be perceived from this description that the course of the current from one of the electrodes is through the sliding rod and the bottom cap, and, therefore, when the rod is drawn up so as not to be in contact with the bottom cap, the current has to pass through the water that intervenes between the end of the rod and the cap. The water being an imperfect conductor of the current, a certain amount of resistance is offered to its passage, and the force of the shock thereby lessened. By drawing the rod further away from the cap, a greater depth of water has to be traversed, and thus the shock can be modified to any required degree.

150. Alternate Character of the Secondary Shock.—It has been previously shown (**36**, **37**) that the induced currents, on completing and breaking the battery circuit, flow in opposite directions, and consequently the shock obtained from the secondary coil is not a uni-direction shock (**148**), but a bi-direction. The current consequent upon *completing* the battery circuit is less in force than that set up when the battery circuit is *interrupted* (**37**).

THE ELECTRIC LIGHT.

151. During the last few years considerable interest has been aroused in Europe and America by the advances in one branch of Electrical Science, viz., that of Electric Lighting. In 1877 so general was the belief that gas would be speedily superseded that a panic occurred amongst the holders of gas shares, resulting in a serious temporary depreciation in the value of these shares.

Sir Humphrey Davy was the discoverer of the Electric Light in the early part of the present century; and he gives, in his "Elements of Chemical Philosophy," a very interesting account of the apparatus he employed. The battery, he tells us, consisted of 2,000 plates, and to its two terminals he attached two wires armed with charcoal points. When these points were brought into contact, a bright spark was emitted, and by withdrawing them from each other to a distance of four inches, a constant electrical discharge took place, forming a most brilliant arc of light. Any substance introduced into this arc was instantly ignited ; platina melted like wax, and quartz, magnesia, and lime all entered into a state of fusion. On more closely examining this arc, Davy found that there was a constant transference of particles of carbon from one electrode to the other, the greater transference being from the negative electrode ; and, further, that the light was due to the incandescence of these particles of carbon, owing to the intense heat of the electric current.

From the time of Davy to the present day, unceasing efforts have been made to develop this beautiful light, and a great advance was made by the invention of Grove's and Bunsen's batteries, for Davy's battery was quite unsuited for general use.

One serious difficulty encountered was the irregular wasting of the carbon points; and many lamps or holders have been designed for maintaining them in one uniform position.

The Electric Lamp in its simplest form, *Fig.* 151, consists of

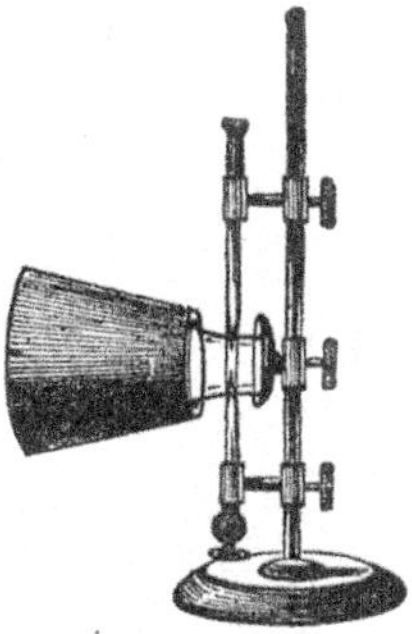

Fig. 151.

two metal sockets for holding the carbon points, placed either vertically or horizontally, with their ends opposed to each other. In addition to this, a means has to be provided for approximating the points ; for, in order to establish the light, the carbons must first touch, and afterwards be separated until the greatest light is obtained.

One of the best forms of Electric Lamps is that designed by M. Dubosq, of Paris.

152. Dubosq's Electric Lamp.—This extremely ingenious apparatus, *Fig.* 152, may be described as a cylindrical frame, supported on a pillar and foot. A central stem rises from the top of the frame, and another longer stem, B, rises also out of the top of the frame, but nearer to its margin and carries a bar, C, at right angles to itself. The pillar, L, supporting the frame is an electro-magnet (**11**), the coils of wire of which form a portion of the battery circuit. The frame contains a horizontal axis carrying two toothed wheels, F, differing in diameter. One of these (the larger of the two) works in a long rack that forms the end of the long stem, B, before described, and the other wheel (the smaller of the two), works in a similar rack that forms the end of the stem that rises out of the *centre* of the frame, the wheels being so placed that the two racks work on their opposite edges.

A carbon point is put into the *central* stem, and another fixed to the under side of the right angle bar, C, that projects from the top of the *side* stem, B, the carbons being so arranged that their points come exactly opposite to each other. When this has been done, the SIDE stem, B, will descend by its own weight, and approximate its carbon towards that fixed to the CENTRAL stem. But in so doing, it turns round the wheels just described, and by this means the carbon in the central stem is *raised* up to meet the descending carbon. In consequence of the dissimilarity in the diameter of these wheels, the central stem does not proceed upwards so fast as the side stem descends, and this difference is arranged so as to compensate for the inequality of rate at which the two carbons waste away.

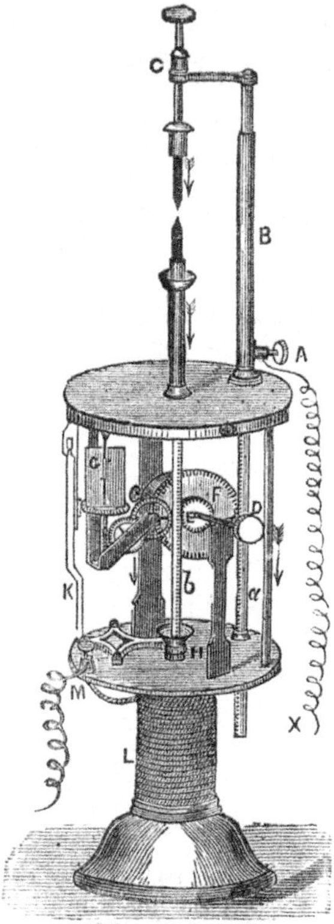

Fig. 152.

When the points of the carbons come into contact all further movement is prevented, one point resting on the other. The lamp is now ready to be attached to the battery, which is effected by fixing the electrodes of the battery to two binding screws, A and M, attached to the lamp. The current passes up the side stem through the two carbons, in the direction of the two arrows, and the central stem, to the wire coiled round the iron bar forming the pillar (converting this into an electro-magnet), and so to the cathode of the battery, and the carbon points will become intensely incandescent. The horizontal axis, carrying the two toothed wheels, F, is now to be turned by means of a milled head, D, on the outside of the cylinder, which will separate the carbon points, and this is to be continued until the maximum light is obtained. Associated with the two toothed wheels on the horizontal axis, before described, is a small train of wheels terminating in an escapement, the pallet of which is attached to a vertical bar, rising up from the bottom of the frame. The end of this pallet bar (as it may be termed) is connected with one end of the lever, the other end of which holds an iron plate coming immediately over the end of the iron bar of the electro-magnet.

The use of this arrangement is to compensate for the wasting away of the carbon points after the battery has been connected, and the proper distance between them adjusted. This result is thus accomplished : the iron bar is formed into a magnet as soon as the battery current circulates round it, and this draws the end of the lever carrying the iron plate downwards, and thus moves, by its opposite end, the pallet bar, which is turned round so that the pallet comes in between the teeth of the escapement wheel, and all further movements of the points is arrested. But when the points waste away the distance between them becomes greater, and the battery current is thereby impeded. This lessens the attractive force of the electro-magnet, and it ceases to hold down the iron at the end of the lever. This movement of the lever acts also on the pallet bar, and carries the pallet out of the escapement wheel, the train of wheels is released, and the carbons gradually approach each other. When these have come near enough to allow the battery current to pass in full force, the power of the electro-magnet is restored, the end of the lever is again drawn down, and this moves the pallet into the escapement wheel, and all further movement is stopped ; this being repeated again and again as the points wear away.

An improvement has been made in this lamp by M. Serrin, of Paris ; and these two instruments, until recently, have held the first place in public estimation.

153. Improved Carbon Points.—It is being realised that further improvement in the mechanical appliances for holding the carbon was impracticable, greater attention has been turned to improving the quality of the carbon points.

Very soon after the discovery of the electric light, it was perceived that wood charcoal was far too soft a material for the points, and an excellent substitute was found in the carbon which is deposited in the necks of the retorts used for the manufacture of gas.

Gas carbon had one serious drawback—its texture is very irregular, and the light is frequently interrupted by pieces breaking off. This defect has been remedied by the manufacture, both in England and in France, of an improved variety of carbon, composed of powdered coke, lamp black, cane sugar, and gum. These materials, by the combined action of the heat and pressure, are made to assume the appearance of cast iron rods, and are in every way well suited for the purpose for which they are intended.

154. Dynamo Electric Machines—The difficulties we have hitherto referred to as preventing the general use of the electric light, have had to do either with the carbon points or the arrangements required for holding the same, but these have not been the sole, nor, in some respects, the most serious impediments. The great want has been a cheap and simple source of electricity; for the employment of 50 or 60 cells of Grove's or any other form of galvanic battery, is a troublesome and costly method of obtaining the electric light.

Thanks to the efforts of the many eminent men who have been investigating the subject, a cheap source of electricity has been found, and the production of the electric light without the aid of a galvanic battery is no longer a dream but a reality.

The machines which are now generally used for the production of the electric light are the outcome of Faraday's discovery of magnetically induced electricity, which has already been described (**12**). It will be remembered that, when a magnet was brought near to a helix of wire attached to a galvanometer, the needle of the galvanometer was instantly deflected, showing that a current of electricity had been induced in the helix by the proximity of the magnet. The electricity obtained in this way is commonly spoken of as magnetic electricity.

The first really successful magneto machine was constructed by Saxton, and afterwards improved by Clark. This consisted of a large permanent horseshoe magnet and a small electro-magnet, the latter being made to revolve in front of the poles of the horseshoe magnet. The wire on the poles of the electro-magnet is wound in suitable directions; and by means of a commutator affixed to its axis, the current obtained flows in one direction only.

The current which this machine gives, while admirably suited for medical and other uses, is not of sufficient power to be utilized for the production of the electric light.

The machines employed for the production of the electric light are known as "Dynamo Electric Machines," and they differ essentially from the "Electric Magnetic Machines," by the substitution of an electro-magnet for a permanent magnet. At the present time there are two or three forms of these machines competing for public favour—the best forms are those invented by Gramme and Siemens.

155. The Gramme Dynamo Machine.—The distinctive peculiarity of the Gramme Machine, *Fig.* 155, is the

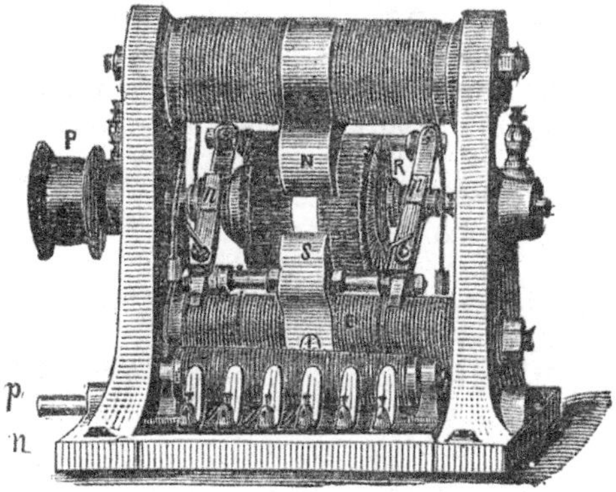

Fig. 155.

construction of the electro-magnet from which the current of electricity is afterwards to oe obtained : this consists of a ring composed of a series of soft iron wires, round which insulated copper wire is wound in distinct coils. The two ends of each coil of copper wire are attached to copper plates, from which the current has to be collected. This ring is made to rotate between the poles of two large electro-magnets which possess a residuum of permanent magnetism, and a part of the current excited in the rotating ring is sent round the electro-magnets to excite them still more. The result of this action is the production of a most powerful electric current, admirably suited for the electric light. The construction of this form of machine will be better understood by reference to *Fig.* 155 ; E, E_1, the two horizontal bars at the top and bottom of the machine, are the two electro-magnets of which N and S are the two poles. Between the two magnetic poles, the Gramme ring, R, rotates, and the current excited in it is gathered up by two friction plates, *p. n.*

In a more recent form of this machine, the Gramme ring is stationary, and the electro-magnets revolve inside it. The ring is composed of soft iron wires, having on it eight distinct coils of insulated copper, each coil being wound on alternately to the right and left. Inside the ring a shaft rotates, carrying eight electro-magnets affixed to it like the spokes of a wheel. The wire on the electro-magnets is wound alternately in opposite directions. The electric current which is needed to excite the electro-magnets is obtained from another machine of a more simple character.

By this arrangement powerful alternating currents of electricity are induced in the coils on the ring, which by suitable connections, are made available for the production of the electric light.

156. The Siemens' Machine, which is the form employed at the Lizard Head Lighthouse, differs considerably from the ones referred to above; it may be roughly described as consisting of a pair of electro-magnets, the iron cores of which are so curved in the centre that, when in position one under the other, there is a circular opening between them in which the armature revolves. The insulated copper wire surrounds the ends of the iron cores, but the centre portion is left free.

The armature is of very unusual form—it is constructed like an induction coil—but with the wire wound on like cotton on a shuttle, instead of after the manner of cotton on a reel. The advantages claimed for Siemens' machines are their diminished size, increased power, and lessened cost.

157. The Brush Machine is the form which has been adopted at the South Kensington Museum, and has also had a very successful trial in the Royal Navy. The apparatus is very simple, consisting of eight electro-magnets recessed on an armature, in which the current is excited by four powerful electro-magnets, and it possesses the great advantage of not requiring a second machine to excite it.

In addition to these three machines there is the Lontin machine, which was used for a considerable time in London for a display of electric light outside the Gaiety Theatre; and the Wallace-Farmer machine, which had an extensive trial at the Liverpool Street Station of the Great Eastern Railway.

158. The Jablochoff Candle.—By the invention of the Jablochoff Candle, *Fig.* 158, Electric lighting has started in a fresh direction, inasmuch as it dispenses with costly lamps, such as have been described.

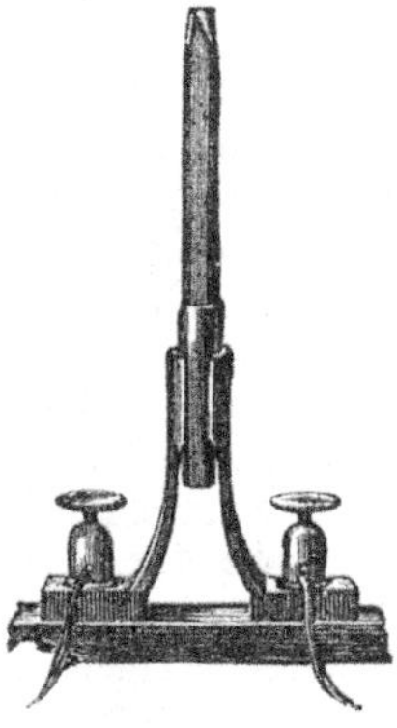

Fig. 158.

In Jablochoff's system, instead of two carbons being placed with their points opposed, they are placed side by side, and are separated or insulated one from another by a strip of non-conducting material, such as plaster of Paris. In order to get over the difficulty of establishing contact between the two points (which, it will be remembered, is a necessary condition before the light can be obtained) Jablochoff connects them together at the top by a narrow bridge of carbon, which is consumed immediately the current is established. In order that the carbons shall be destroyed at an equal rate, an alternating current of electricity is necessary, such as is supplied by the Gramme machine; the insulating material being simultaneously fused by the intense heat of the electric arc.

Then one difficulty associated with the Jablochoff Candle is its rapid destruction, a candle not lasting more than one and a half hours. This difficulty is overcome by grouping three or four candles together; and, when one candle is consumed, by an ingenious appliance called an "Automatic Shunt," the current is diverted to the next one, and so on. The Jablochoff system was employed for illuminating the Place de l'Opera, Paris, during the late Exhibition, and has likewise had an extended trial on the Thames Embankment.

159. Various Other Systems of Electric Lighting.—In addition to the Jablochoff system several others have been attracting attention; there is the Wallace-Farmer Regulator, the Voltaic Arc of which is obtained from the edges of two carbon plates, 9 inches long and 3 inches broad, the upper plate being about twice the thickness of the lower one, and their position regulated by magnetic contrivances.

Then, again, there is the system of M. Rapieff, which has been adopted at *The Times* printing office, where each electrode is composed of two carbons, each couple being arranged in the shape of the letter **V**. This system requires a special form of lamp or regulator.

Then, again, there is the Werdermann regulator, in which the lower carbon pencil is made to impinge on to a disc of carbon; the pencil being kept pointed by the action of the current.

Of all inventors whose attempts to solve the problem of electric lighting that have been made public, none have attracted so much attention as Mr. Edison, of New York, and his efforts have chiefly been to find some imperishable material which should itself become entirely incandescent by the passage of the electric current. Edison's efforts have not hitherto been crowned with success, the materials employed being sooner or later destroyed. The efforts of many other inventors in this line have likewise terminated unsuccessfully.

ELECTRIC BELLS.

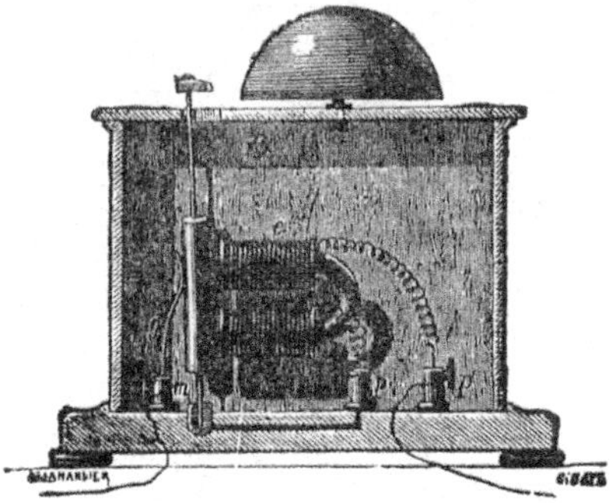

Fig. 160.

160. The Construction of an Electric Bell.—The employment of the electric current for ringing bells has become general, both in business and private establishments.

The mechanism of an ordinary Electric Bell, *Fig.* 160, is very simple; it is, in fact, a "Vibrating Contact Breaker," (**51**), to the spring of which a small hammer is affixed, so that when the spring vibrates, the hammer shall strike a bell fixed within its reach.

161. The Best Form of Battery to use for an Electric Bell is the Leclanché (**161**). When the circuit through which the current has to pass is comparatively short, two Leclanché cells will suffice; but if the circuit be a long one, from four to six cells will be required. The battery should be placed in a cellar, or other place, where the temperature is uniform.

162. The Wire Used for the circuit should be the No. 20 copper wire, and it should be covered with gutta-percha and cotton. As far as possible, the wire should be laid in dry situations, and great care should be taken that insulation of the wire should be in no way impaired.

If the wire has to be carried overhead for long distances, ordinary galvanised iron wire may be used, the gauge varying from No. 4 to No. 14, according to the length of line. The wire, under these circumstances, has to be supported on poles, by insulators; care, of course, being taken that the wire does not come into contact with the branches of trees, &c., which would cause a leakage of the current.

163. The Press Button.—In addition to the battery, bell, and wire, a ready means is required for instantaneously completing and interrupting the circuit; this is accomplished by what is called a "Press Button." The press button, *Fig.* 163, is a cylindrical box, in the centre of which is an ivory button attached to a brass spring; by pressing this button the spring is brought into contact with a brass plate on the bottom of the box. One wire of the battery is attached to the spring of the press button, and one wire from the bell is attached to the brass plate in the bottom of the box. It will be seen that when the button is in its normal condition, the circuit between the bell and the battery is interrupted; but when the button is pressed down, the spring to which it is attached is brought into contact with the brass plate, and thus the circuit is completed and the bell made to ring.

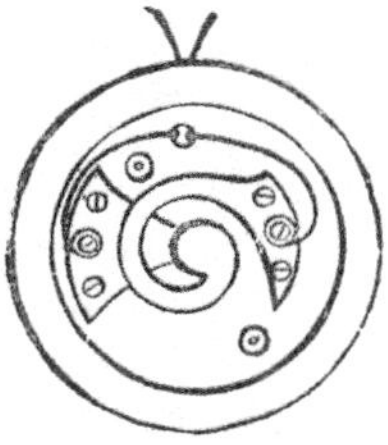

Fig. 163.

164. The Method of connecting the various parts is shown, *Fig.* 164. A is the Leclanché cell, B the press button, C the bell, and D the return wire.

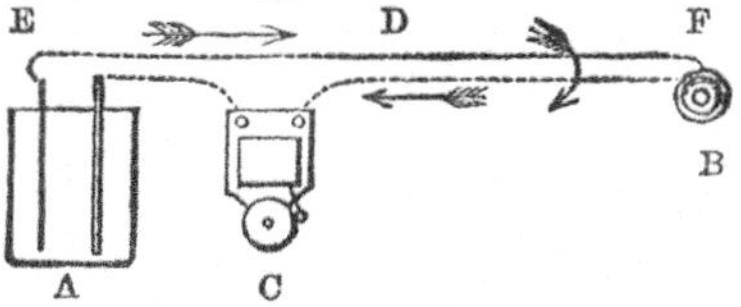

Fig. 164.

165. The Earth Current.—When the distance between the bell and the button is very considerable (half a mile, for instance), the return wire, D, *Fig.* 164, may be dispensed with, and the earth used as a substitute by attaching terminals, E and F, to copper plates sunk in the ground. This valuable discovery was made by Professor Steinheil in 1837, and is now universally adopted.

THE ELECTRIC TELEGRAPH.

166. The utilization of the electric current for instantaneously sending messages to distant places is one of the most valuable inventions of modern science. The germ of the Electric Telegraph is found in Oerstead's discovery (**6**) of the power which an electrified wire has of deflecting a magnetic needle; the deflection being to the right or the left, according to the direction in which the current flows.

167. Wheatstone's Telegraph.—One of the earliest forms of the Electric Telegraph used in England was invented by Prof, Wheatstone in 1837, and its principle is that of the galvanometer (**7**). The needle is made to work vertically, in front of a card, on which is inscribed the letters of the alphabet, and their corresponding telegraphic signs. The needle is made to deflect to the right or left by changing the direction of the current, this being effected by means of a commutator or key.

168. Dial Telegraphs.—For those who do not care for the trouble of learning the telegraphic signs, there are the dial telegraphs. In these instruments the letters of the alphabet are arranged on a ring, round the front of which a pointer is made to move, letter by letter, by the aid of an electro-magnet. The transmitter used for sending the message is a dial with letters in a circle, over which a lever with a handle is moved; as the lever is moved round, the current is interrupted at every other letter. This, of course, affects the electro-magnet at the other dial, and the movement of the attached keeper, by a simple mechanism, is communicated to the pointer, whose movements correspond to the transmitter. The messages are spelt out letter by letter.

169. The Morse Alphabet.—On the Continent the Morse Telegraph is the one most in use, and its peculiarity consists in its making an automatic record of the message sent on a strip of paper. The paper is unwound from a reel, and made by clock-work to pass in front of a pencil. The movements of the pencil are controlled by an electro-magnet; and, according to the length of time the pencil is pressed against the paper, a dot or a dash is inscribed thereon.

170. The Morse Alphabet.—The following is the Morse Alphabet :—

A . —	K — . —	U . . —
B — . . .	L . — . .	V . . . —
C — . — .	M — —	W . — —
D — . .	N — .	X — . . —
E .	O — — —	Y — . — —
F . . — .	P . — — .	Z — — . .
G — — .	Q — — . —	Full stop . . . . . .
H	R . — .	Repetition . . — — . .
I . .	S . . .	Hyphen — . . . . —
J . — — —	T —	Apostrophe . — — — — .

1 . — — — —	6 — . . . .
2 . . — — —	7 — — . . .
3 . . . — —	8 — — — . .
4 . . . . —	9 — — — — .
5	0 — — — — —

In fixing wire for the Telegraph, the arrangement described under the heading of "Electric Bells" should be followed.

THE TELEPHONE.

171. The application of electricity for the reproduction of the human voice has opened up a whole sphere of usefulness; and, since the invention of the Telephone, it is possible not only to hear the words uttered by a distant friend, but also to recognise the very tones of his voice.

172. The Transmission of Sound.—In order to understand the principle of the Telephone, a brief description of the way in which sounds are transmitted may be necessary.

That peculiar sensation in the organ of hearing which is called "a-sound" is due to vibrations or waves in the air acting upon the tympanum of the ear. Thus we strike a blow on a drum; not only is the parchment of the drum made to vibrate, but similar vibrations are set up in the surrounding air, and thus bystanders become conscious of the sound. It will be therefore perceived that in the transmission of sound the particles of air are not transported from place to place, but the vibration in one particle is communicated to the next, and so on, the intensity of the vibrations becoming less and less as the distance from the original cause of the disturbance increases.

All sounds need for their transmission some elastic medium, such as air, wood, or metals, and if these be not present, the drum may be beaten ever so hard yet not a sound would be heard. In all sound transmitters this fact has to be recognised; thus, in the speaking tubes which were so common before the introduction of electric bells, the sound of the voice is conveyed to great distances by confining the vibrations in the narrow column of air enclosed in the tube.

173. The Phantom Orchestra.—Some years ago a great sensation was made at the "Polytechnic" by an orchestra of instruments giving out their respective musical sounds without any players being present. In this case the musicians were playing on another set of instruments in the basement of the building, and the sounds were conveyed to the instruments in the room above by means of wooden rods.

174. The Lover's Telegraph is another adaptation of this principle. It consists of two open wooden or tin boxes, connected together by a string passing through the centre of the bottom of each. The words spoken into one box are distinctly heard by applying the ear to the other box (the string, of course, being stretched tight). The elastic medium which conveys the sound in these illustrative cases is respectively,—the air enclosed in the speaking tube,—the wooden rods,—and the string.

175. Noise and Music.—It may be well to explain that the difference between a noise and a musical sound is, that the vibrations in the former case are irregular in their intervals, whereas, in the latter case, the vibrations occur at regular intervals. The greater the number of vibrations per second the higher the note produced; thus, to produce the sound corresponding to the lowest C in the bass 120 vibrations per second are needed, but to produce the C an octave higher 256 vibrations are needed. This fact supplies an explanation of the modulation we observe in the human voice, the air being caused to vibrate at different rates by the act of speaking.

176. The Telephone differs from the instruments just described in this respect, that whereas the air, the rods, and the string *conveyed* the vibrations, and can be used only over limited distances, the Telephone *reproduces* vibrations, and can do so at practically unlimited distances.

177. Bell's Telephone, *Fig.* 177, which is the one most

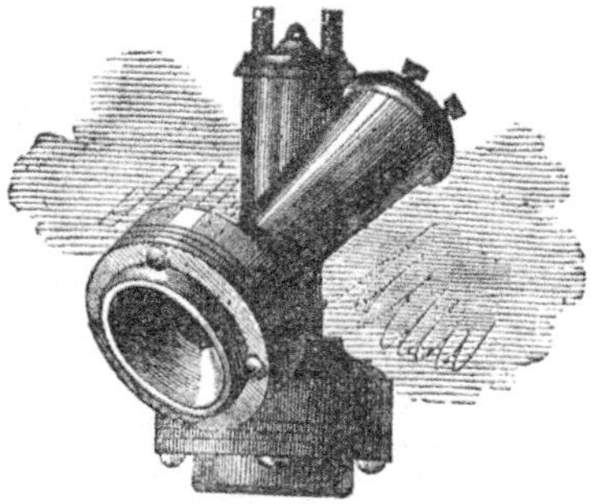

Fig. 177.

generally known in England, consists of a bar magnet, about 5 inches long and $\frac{1}{4}$ inch in diameter, on one end of which is fitted a wooden bobbin, wound with No. 38 silk-covered wire. The magnet and bobbin are fitted into a trumpet-shaped wooden case, which is covered over at its broad end with a thin iron plate (such as photographers use for ferrotype portraits), the iron plate and the magnet almost touching each other. The iron plate is kept in its position by screwing over it a wooden disc, shaped like the mouthpiece of a speaking tube, and having a hole in the centre. Any vibrations which the iron plate may experience (such as would be caused by speaking on to it) disturbs the electrical condition of the wire round the end of the magnet, and if the two wires of the Telephone be joined in circuit with the two wires of a similar instrument, the electrical disturbance in the coils of the first will be communicated to the coils of the second, causing thereby the attached iron plate to vibrate and reproduce the original sound. It is manifest that these sounds are *reproduced* by the Telephone, and not *conveyed* by the wire, from the fact that sound travels at the rate of 1,100 feet per second, but the Telephone reproduces sound instantaneously.

178. Reis' Telephone.--Prof. Bell is not by any means the only person investigating, or even the first investigator, in this branch of electrical science. To Reis, of Homburg, the credit of producing the first Telephone must be given. The principle of Reis' Telephone is the making and breaking of an electric circuit by the vibration of a tightly-stretched piece of bladder against a strip of platinum; the vibrations are reproduced by a similar instrument at the other end of the circuit.

179. Dolbear's Telephone.—Prof. Dolbear, of Massachusetts, U.S.A., is the inventor of an instrument similar in principle to that of Bell, the chief difference being the substitution of a horseshoe magnet for the bar magnet.

One of the early difficulties experienced with the Telephone was that sounds were reproduced only in whispers, and Mr. Gower is the inventor of a more audible Telephone. In Gower's Telephone a horseshoe magnet is employed, with coils of wire round its two poles, the vibrating diaphragm is of increased size, and is mounted on a kind of sounding board.

180. Edison's Telephone.—Mr. Edison has invented a loud speaking Telephone, the principle of which is electro-chemical. The instrument consists of three parts—a battery, a sound receiver, and a sound reproducer. In the receiver, the sound waves impinge on a wafer of smoke carbon, through which the current is flowing to the wire, which constitutes the circuit, and by the vibrations set up in the carbon the strength of the current is regulated. The sound reproducer consists of a diaphragm of mica, standing out from the centre of which is a metal spring, shod at its free end with a piece of platinum. This spring rests on a rotating cylinder of chalk which is kept moistened with a solution of caustic potassa and acetate of mercury. Every time the electric current passes from the spring to the chalk a diminution in the amount of friction takes place proportional to the strength of the current, and thus the mica diaphragm is thrown into vibration. It will be seen that, in this instrument, the real source of power is the battery, hence the sounds reproduced may be stronger even than those received.

THE MICROPHONE.

181. The Microphone.—Reference has been made to the fact that elastic bodies, such as the air, wood, and metals, are good conductors of sound. Prof. Wheatstone in 1840, while perfecting his electric clock, made a very interesting discovery, that an electrified wire (*i.e.*, a wire along which an electric current was passing) would convey sounds to a much greater distance than a non-electrified wire—the sounds of a ticking clock being distinctly audible through several miles of wire.

This subject has been recently still further investigated by Prof. Hughes, who, in May, 1878, communicated to the Royal Society his remarkable discovery of the Sound Magnifier, or Microphone.

A Microphone consists of a cylindrical piece of gas carbon, pointed at each end, about one inch long, and one-eighth of an inch in diameter at the centre, supported on a wooden stand between two pieces of carbon, about one inch square, to which are attached two brass winding screws. The carbon is made a better conductor of electricity by being made red hot, and then plunged into a dish containing mercury.

182. To Use the Microphone a small battery is required (a Smee or Daniel will do), one pole of which is attached to one of the screws of the Microphone, another wire passing from the second screw of the Microphone to one of the terminals of a telephone, the circuit being completed by joining the other pole of the battery to the second terminal of the telephone. The Microphone acts as the recipient of the sounds, the telephone reproduces them.

By this contrivance sounds, which ordinarily cannot be distinguished, become distinctly audible the tick of a watch, the beating of a pulse, the tramp of a fly being heard at least a hundred miles distant from the source of sound.

Prof. Hughes, in his explanation of the cause of this remarkable increase of sound, speaking of the action of the Microphone says:—"It is quite evident that these effects are due to a difference of pressure at the different points of contact, and they are dependent for the perfection of action upon the number of points of contact." Sounds are conducted by the molecules of matter, and in the Microphone we have the mercury in the carbon in a finely-divided state. The effect of sound upon these finely-divided particles of matter is to alter their dimensions, bringing more or less of them into contact; thus, the resistance which the mass of a conductor offers to the passage of the electrical current continually varies, and, therefore, the effect of the current upon the telephone is intensified.

183. Preece's Telephone.—At a meeting of the Royal Society, in May, 1880, Mr. Preece, of the London Post Office, read a paper on "Some Thermal Effects of Electric Currents," in which he showed that extremely thin wires acquire and lose heat so rapidly, and are so sensitive to linear expansion that, if inserted in the circuit of a Microphone transmitter, they reproduce sonorous vibrations. Thus, a mere sounding board or disc and a fine wire are all that are needed to reproduce speech with the same faithfulness that Bell obtains with electro-magnetic agency, Edison with electro-chemical agency, and Hughes with his Microphone principle. The Preece telephone, if it do not prove so practical as other forms in use, has certainly carried simplicity of construction to its utmost limit.

THE PHONOGRAPH.

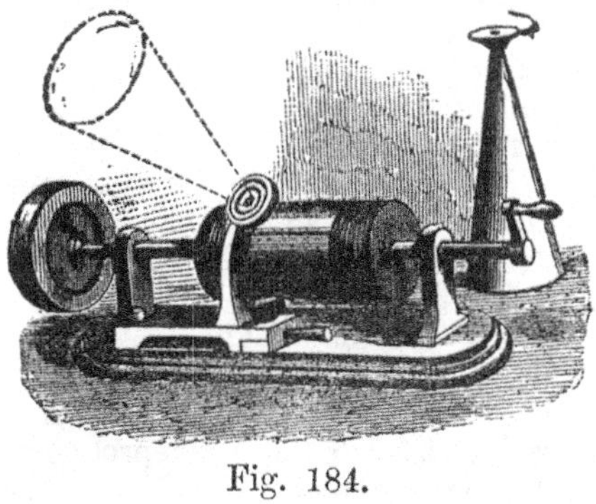

Fig. 184.

184. The Phonograph, *Fig.* 184, though not exactly belonging to our subject, ought perhaps to receive a brief descripton, inasmuch as its invention is an outcome of the investigation of the phenomena connected with the telephone.

The Phonograph (an invention of Mr. Edison, of New York) is an instrument by which sounds can be imprinted on soft metal, such as tin-foil, and by it reproduced with distinctness and accuracy of tone at will any number of times.

This remarkable instrument consists of three parts—a receiver, a recorder, and a reproducer.

The receiver consists of a curved tube, one end of which is fitted with a mouthpiece, and the other end closed by an exceedingly thin metal plate or diaphragm, which vibrates with the least sound. On the centre of the outer side of this metal diaphragm is a small blunt steel pin. The recorder is a brass cylinder 4 inches long and 4 inches in diameter, with a continuous V-shaped groove cut into it from end to end, like a large screw. Two pins 4 inches long are fixed in the ends of the cylinder, one of which is cut with a screw thread, and these pins are fitted into appropriate bearings, eight inches apart. By turning a handle, fixed on to one of the pins just referred to, the cylinder is not only rotated, but traverses from one support to the other.

In using this instrument the recorder is covered with a sheet of tin-foil and is placed in front of the vibrating diaphragm of the receiver, the blunt pin of which just touches the surface of the tin-foil, and is opposite the commencement of the groove of the brass cylinder. If, now, a person speaks into the mouthpiece of the receiver, and the handle of the recorder be turned, a spiral series of indentations will be made on the tin-foil by the pin of the vibrating diaphragm, these indentations corresponding to the groove in the brass cylinder under the tin-foil. The sounds that have been uttered are now recorded on the tin-foil.

The reproducer, which forms the third part of the Phonograph, consists of a conical metal drum, having its larger end open; the smaller, which is about two inches in diameter, being covered by a sheet of paper stretched tight like a drum. In front of this paper diaphragm is a light flat steel spring, held in a vertical position, and from the end of which a blunt steel point projects. This spring is connected with the paper diaphragm by means of a silken thread, which is placed just sufficiently in tension to cause the other face of the diaphragm to assume a slight convex form.

In order to reproduce the sounds received on the tin-foil, the receiver is placed in front of the "reproducer," so that the blunt pin is just over the first indentation; the handle of the receiver is turned in a reverse direction to what it was before, and this will cause the spring of the reproducer to vibrate. The movements of the spring are communicated to the paper diaphragm by the silken thread, and the words spoken into the receiver will now be heard issuing from the open end of the reproducer.

Since the Phonograph was first brought out several improvements have been made, the most important of which is fitting a clock movement to the receiver to rotate it at a uniform speed.

ELECTRO MOTORS.

185. The application of electricity as a motive power is, at the present time, one of the most promising fields for investigation. That electricity is a source of power is demonstrated in the Vibrating Contact Breaker (*Fig.* 51), where an electro-magnet attracts a piece of iron every time the current flows. All Electro-Motor machines are constructed on the principle of alternate making and breaking battery contact; and in one of the most ordinary forms of machine a wheel is constructed, something like the paddle-wheel of a steamer with pieces of soft iron substituted for the paddle board. This wheel rotates in front of four electro-magnets, the electric current being supplied by an ordinary battery. The passage of the current is regulated by a toothed wheel contract breaker attached to the axis of the large wheel just referred to, so that every time a soft iron bar is attached opposite to the poles of an electro-magnet the current is shut off, and the impetus which the wheel thus obtained being sufficent to carry it forward and subject another piece of iron to the influence of the electro-magnets.

At the Berlin Industrial Exhibition of 1879, Dr. Werner Siemens fitted up an electric tramway, and during the course of the summer upwards of 100,000 persons practically tested it.

The source of power was two dynamo-electric machines, one being stationary, and the current generated by it being conveyed to the second machine, which was mounted on wheels in front of the train. By these currents an armature is made to revolve, and this motion is in turn communicated to the wheels of the car. The conductors of the currrent were the rails upon which the carriages travelled.

At the Agricultural Show at Bar-le-Duc, France, in May, 1880, an Electro-Motive machine was exhibited suitable for agricultural use. A large field was most successfully ploughed with one of Howard's double-furrow ploughs worked by this new machine.

SEEING BY ELECTRICIT

186. A new electrical wonder has just been announced, viz., "Seeing by Electricity," and a patent has just been taken out for the instrument by the inventor of the telephone. The character of this instrument has not as yet been made public, but Messrs. Perry and Ayrton have communicated the following facts to *The Times*, which English electricians have independently discovered, and these tend to show that "Seeing by Electricity" must not be regarded as an impossibility:

"Our transmitter at A—that is, the apparatus for receiving the light impressions and transmitting them electrically—consisted of a large surface made up of very small separate squares of selenium. One end of each piece was connected by an insulated wire to the distant place, and the other end of each piece with the ground, in accordance with the plan commonly employed with telegraph instruments. The object whose image was to be sent by telegraph was illuminated very strongly, and by means of a lens a very large image was thrown on the surface of the above transmitter. Now, it is well known if each little piece of selenium forms part of a circuit in which there is a constant electro-motive force, say of a voltaic battery, the current passing through each piece will depend on its illumination. Hence the strength of the electric current in each telegraph line would depend on the illumination of its extremity. Our receiver at the other end, B, was, in our original plan, a collection of magnetic needles the movements of each of which (as in the ordinary needle telegraph) were controlled by the electric current passing through the particular telegraph wire with which it was in connection. Each magnet by its movement closed or opened an aperture through which light passed to illuminate the back of a sheet of frosted glass. There were, of course, as many of the illuminated squares at B as of selenium squares at A, and it is quite evident that since the illumination of each receiving square depends on the strength of the current in its circuit, and this current again depends on the illumination of the selenium at the other end of the wire, the image of a distant object might in this way be transmitted as a mosaic by electricity.

"A more promising arrangement, suggested by Professor Kerr's experiments, consisted in having each little square at B made of silvered soft iron, and forming the end of the core of a little electro-magnet, round which passed the current, coming from the corresponding selenium square at the other end. We proposed that the surface formed by these squares at B, should be illuminated by a great beam of light, *polarised* by reflection from glass, and received again by an analyser. It is then evident that since the intensity of the analysed light depends on the rotation of the plane of polarisation by each little square of iron, and this depends on the strength of the current, and that again on the illumination of the selenium, we have another method of receiving at B the illumination of the little squares at A. It is probable that the sealed description of Professor Graham Bell's may relate to some plan of a very much simpler kind than either of ours; but as the two we have mentiond are the outcome of the electrical progress England has made during the last few years, it may be desirable to put them on record."

PERKEN, SON & RAYMENT'S

OPTICAL

AND

SCIENTIFIC INSTRUMENTS.

SPECTACLES, Eye Glasses, Folders, Hand Frames, Lenses either Spherical, Cylindrical, Sphero - Cylindrical, or Prismatic, White or Coloured Glasses of all Foci.

OPERA, Field, or Marine Glasses of every description.

TELESCOPES, Binocular of Highest Power.

TELESCOPES, Monocular, Powerful.

MICROSCOPES, Microscopic Objects, Cabinets, Lamps, and all Accessories.

THERMOMETERS, Clinical, Chemical, Air, Sixes, Maximum and Minimum, &c., &c.

BAROMETERS, Mercurial, Board of Trade, Marine, Pit, & Household.

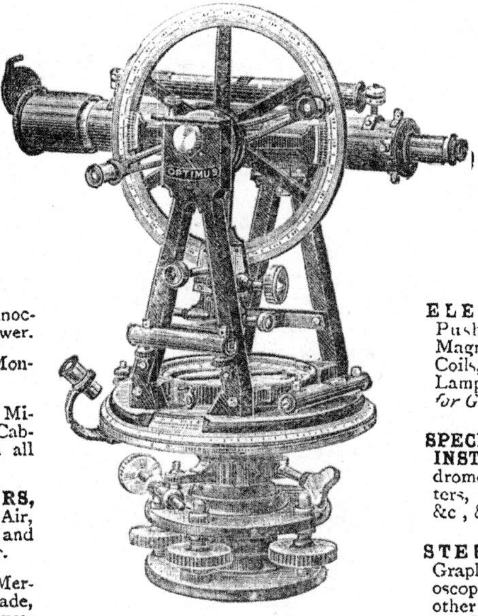

ANEROIDS Ships' Clocks, Sextants, Compasses.

MATHEMATICAL & SURVEYING INSTRUMENTS, Theodolites, Rules, Scales, T-Squares, Set Squares Curves, Spirit Levels, Compasses.

ELECTRIC Bells, Pushes, Batteries, Magnetic Machines, Coils, Galvanometers, Lamps, Wire. *Agents for Gaiffe, Paris.*

SPECIFIC GRAVITY INSTRUMENTS, Hydrometers, Salinometers, Saccharometers, &c , &c

STEREOSCOPES, Graphoscopes, Praxinoscopes, Reading and other Magnifiers, Camera Obscura.

TRADE DISCOUNT LIST ON APPLICATION.

SPECTACLES, EYE GLASSES, &c., &c.

Convex or Concave Glasses, blue steel spectacle frames, from 5s. to 42s. per doz.
" " Grooved Glasses, curl-side steel spectacle frames, from 12s. to 42s. per doz.
" " Glasses, gold spectacle frames, per pair, 12s. 6d., 17s. 6d., 25s., 30s., 40s.
Double Eye-glasses, (folding), steel frames, 9d. to 42s. per pair.
" " tortoise-shell frames, 12s. to 36s. per pair.
" " steel frames, grooved glasses, 12s. to 36s. per pair.
" " gold frames, 12s. 6d., 17s. 6d., 25s., 30s., 40s.
Single " (Oxfords) 3s. to 9s. per doz.
Best Brazilian Pebbles to Spectacles or Eye Glasses, extra, Convex, 24s. to 100s
" " " cut in the axe, extra, 42s. to 72s.
Coloured Spectacle Lenses of various tints, either concave, convex, globular, or parallel.
Ladies' Hand Frames either of Tortoise-shell or Gold, inlaid or jewelled.
Cylindrical, Sphero-cylindrical Spectacles for cases of Astigmatism, specially made to Oculist's formulæ.
Cases, for Oculists, containing Lenses for testing sight (spherical, cylindrical, prismatic) Trial Frames, Stenopaic Discs, &c., &c.

MAGNIFYING GLASSES FOR PICTURES, READING, &c.

Reading and Picture Magnifiers, in shell, ivory and metal, &c., 6d., 9d., 1s., 1s. 6d., 2s. 6d., 3s. 6d., 5s., 10s., 12s. 6d., 15s.
" " " (cylindrical), in shell ivory, metal, &c., 7s. 6d., 10s. 6d., 15s., 20s., 30s. 40s.
High-power Magnifying Glasses for Botanists, in ebonite, shell, or metal 1s., 1s. 6d., 2s. 6d., 3s. 6d., 5s. 10s., 15s.

GRAPHOSCOPE, STEREOSCOPES, &c.

In Black Wood, ornamented with gold or painted flowers, 4s., 5s., 9s., 12s.
Grapho-Stereoscopes combined, 12s., 18s., 24s., 38s.
" Walnut wood, superior, 5in. lens, 48s.; 6in., 60s.
Stereoscopes, 3s. 9d., 5s.; Achromatic, 13s., 15s., 19s., 30s. 40s.
" on pedestal stands, 25s., 30s., 40s. 50s.
" Revolving, for 50 slides, 60s.
" " 100 " 130s.

PEDOMETERS.

	£	s.	d.
In Nickel-plated cases, registering to 12 miles	0	10	0
Ditto, Silver cases (Hall-marked)	1	0	0

OPERA, FIELD AND MARINE GLASSES.

OPERAS, Achromatic, in Cases.

Diam. of Object-glass	1 1-8	1 5-16	1¼	1 5-8	1 7-8
Morocco-cov'd bodies:	*s. d.*	*s. d.*	*s. d.*	*s. d.*	*s. d.*
Japanned mounts...	6 9	7 3	9 0	11 6	13 3
„ better qlty.	15 0	20 0	24 0	28 6	32 0
„ 12 lens, supr.	21 9	27 0	32 0	36 0	42 0
Ivory Bodies:					
Gilt mounts... ...	14 6	18 0	23 0	27 0	42 0
„ 12 lens, supr.	30 0	38 0	45 0	50 0	60 0
Pearl bodies:					
Gilt mounts	30 0	36 0	41 0	48 0	60 0
„ velvet cases, 12 lens, superior ...	57 6	66 0	95 0	120 0	150 0
Aluminium mounts. superior	48 0	60 0	68 0	77 0	90 0
Morocco or Russia	40 0	50 0	57 0	64 0	75 0
Pearl bodies	66 0	79 0	93 0	108 0	130 0
Tortoise-shell bodies	60 0	79 0	94 0	108 0	130 0
Inlaid Pearl, Ivory or Enamelled Bodies, very handsome, ... superior	60 0	90 0	125 0	150 0	180 0

FIELD AND MARINE, Achromatic, in Sling Cases.

Diam. of Object-glass	1 3-8	1 5-8	1¾	2	2¼
Morocco bodies:	*s. d.*	*s. d.*	*s. d.*	*s. d.*	*s. d.*
Japanned mounts ...	15 0	18 0	24 0	28 6	34 0
„ „ better ...		32 0	35 0	38 6	41 0
Morocco or Russia, 12 lens, superior	40 0	44 0	50 0	56 0	60 0
Japann'd mnts., 12 lens, supr.	34 0	36 6	41 6	46 6	50 0
Aluminium, 12 lens, supr.	102 0	144 0	138 0	155 0	175 0

For 2-draw Field or Opera Glasses, add extra 10s to 30s.
For Bending Bar ditto to adjust centres, add extra 12s 6d. to 34s.
Marine Glasses, specially adapted for night use, having large object-glasses and eye-pieces, 40s. to 50s.

BINOCULAR TELESCOPES.

Magnifying from 12 to 25 diameters, or 144 to 576 times.
The definition will be found exceptionally fine.
Arrangement for adjusting width between Eye-pieces.

Diam. of Object-glass.	1	1 1-8	1 3-8	1 5-8	1 7-8
Morocco or Russia bodies: Japanned mounts, with	*s. d.*	*s. d.*	*d.*	*s. d.*	*s. d.*
sun or spray shades ...		125 0	135 0	145 0	200 0
Aluminium mnts., shades ...		200 0	265 0	298 0	415 0
Morocco Conical bodies:					
Japanned mounts, no shades	84 0	92 0	100 0	...	...

The above prices include Sling Case.

TRADE DISCOUNT LIST ON APPLICATION.

'OPTIMUS' DWARF OPERA,

Leather Covered,

15/-

'OPTIMUS' GUINEA OPERA,

Morocco Covered,

21/-

'OPTIMUS' PEARL OPERA,

25/-

'OPTIMUS' ALUMINIUM OPERA.

50/-

'OPTIMUS' ECONOMIC FIELD GLASS.

Small Size ....
Clear Definition ....
Good Field of View Magnifying 16 times.. .... } 21/-

'OPTIMUS' SCORER.

Is unrivalled for excellence and cheapness. It shows the number of people in boats four miles distant, Sea-birds one mile distant, and Bullet-marks on Target at 600 yards.
Magnifying 25 times. } 30/-

'OPTIMUS' SCOUT.

Medium Size ....
Clear Definition ....
Good Field ....
High Magnifying Power (64 times). } 70/-

'OPTIMUS' BINOCULAR OR DOUBLE TELESCOPE.

This Instrument has Bending Bar for Exact Visual Adjustment...
The Size is Moderate ....
The Definition is Crisp .. ..
Very high (144 times) Magnifying } 80/-

If with Shades, as Diagram ......} 100/-

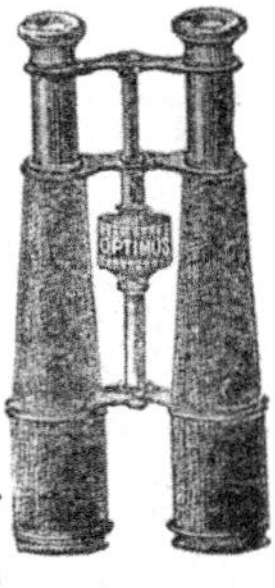

For General Excellence, Definition, and Magnifying Power, we invite intending Purchasers to test our Field and Opera Glasses against any in the world.

Opera Glasses in great variety.

TRADE DISCOUNT LIST ON APPLICATION.

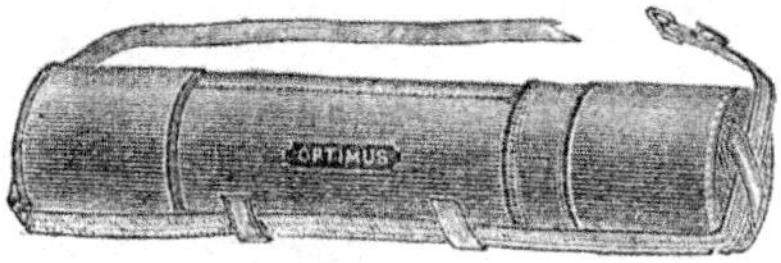

'OPTIMUS' TELESCOPES.

ACHROMATIC.

Diameter of object glass	1in.	$1\frac{3}{16}$	$1\frac{7}{16}$	$1\frac{11}{16}$	2
Morocco covered bodies—	s. d.	s. d.	s. d.	s. d.	s. d
Brass mounts, 3 draws	5 9	8 0	15 0	21 0	
„ „ and spray or sun shades	7 0	10 6	17 6	23 0	
"Tourist" Morocco-covered bodies—					
Brass mounts, 3 draws and cap and sling ...		12 6	19 0	27 0	37 0
Oxidised mounts, 3 draws, cap and sun shades		15 6	23 0	30 0	41 0
Diameter of object glass	$1\frac{1}{16}$	$1\frac{7}{16}$	$1\frac{5}{8}$	2	$2\frac{1}{4}$
"Army" Morocco-covered bodies—	s. d.	s. d.	s. d.	s. d.	s. d.
Oxidised mounts, 4 draws		31 0	41 0	58 0	72 0
"Navy" or "Day and Night" Morocco-covered bodies—					
1 draw, with shades		30 0	41 0	60 0	82 0
"Government," Morocco-covered bodies—					
Pancratic, 3 draws, with shades		42 0	54 0	72 0	90 0
"Yachtsman," Morocco-covered bodies—					
Tapered, 1 draw, with shades		34 0	42 0	57 0	77 0
"Rifleman's," Morocco-covered bodies—					
2 draws, exceptionally portable, but having high power		29 0	34 0	...	...
With still higher power		...	40 0	...	...
"Deerstalker," Morocco-covered bodies—					
3 draws, cap and sling		26 6	...	44 0	58 0
"Viceroy," Morocco-covered bodies, 3 draws ...		25 0	...	40 0	55 0
"Empress," for ladies' use, being very light—					
Polished Aluminium bodies, covered with superior leather, 3 draws ...	80 0	95 0	...	130 0	...

Wood Tripod Stands for Telescopes, with gutter tops, adjustable, from 13s. to 30s.

ASTRONOMICAL.

Diameter of object glass	$1\frac{1}{16}$	$1\frac{7}{16}$	$1\frac{11}{16}$	2
Portable, Morocco-covered, terres and celes eye-pieces	s. d. 9 0	s. d. 18 0	s. d. 23 0	s. d. 36 0
„ on wood tripod stand for table, fitted in cabinet, complete	...	...	34 0	
Diameter of object glass	$2\frac{1}{4}$	$2\frac{1}{2}$	$2\frac{3}{4}$	3
Brass body, best finish, rack motion for focussing, mounted on brass claw stand, vertical and horizontal adjustment, terrestrial and celestial eye-pieces, fitted in cabinet, complete ...	£ s. 9 0	£ s. 10 5	£ s. 14 0	£ s. 16 10
Diameter of object glass	3	$3\frac{1}{2}$	4	$4\frac{1}{4}$
Superior, as above, fitted with finder, rack steadying rods, 2 terrestrial and 2 celestial eye-pieces, on best rigid stand, in cabinet, complete ...	£ s. 30 0	£ s. 35 0	£ s. 54 0	£ s. 70 0

'OPTIMUS' MICROSCOPES.

	£	s.	d.
Youth's, French made with single objective, in wood case	0	6	3
„ Superior, 11s. 3d., 15s. 9d., £1, £1 6s., £1 15s., £2 1s.,	3	2	0
Student's, English made, dividing objectives, giving 3 powers	3	6	6
„ Superior, with rack adjustment, with 2 eye-pieces—2 objectives—condenser, &c., &c., £5 10s., £7 10s.,	10	0	0
University, dividing objective, giving 3 powers ...	3	3	0
Monocular, rack and pinion adjustment—glass stage, reflector, condenser, 2 eye-pieces—2in., 1in. and ¼in. objectives	15	0	0
Monocular, as above, with Polarizing apparatus, 3 eye-pieces—2in., 1in., ¼in. and ⅛in. objectives, double nose-piece...	22	10	0
„ Large size, circular revolving stage, graduated, milled heads to slow motion, suited for examining sections of crystals—3 eye-pieces, 2in., ¼in., ⅛in. and $\frac{1}{10}$in. objectives	38	10	0
Binocular bodies, having the necessary additional eye-pieces suitable for the instrument ... extra from £4 to	9	0	0
Dissecting microscope, complete	2	10	0

Inches	4	3	2	1	1	⅔	½	¼	⅕	⅙	⅛	$\frac{1}{12}$	$\frac{1}{15}$
Microscopic Objectives ...	25	25	25	25	25		40	35	40	48	65		
Ditto, 1st quality—													
Angular aperture ...	9°	12°	16°	20°	27°	30°	80°	100°	100°	150°	140°	145°	150°
Greatest magnifying power ... diameters	72	112	140	196	280	420	620	1400	1480	1820	2380	5200	6400
	26 6	40	40	46	46	50	77	88	105	125	135	198	210

The first quality, half-inch and higher powers, are fitted with screw collar arrangement for adjusting distance according to thickness of covering glass.

	£	s.	d.
Polarising Apparatus ...	1	2	0
Parabola	0	16	0
Spot Lens	0	8	0
Double Nose-piece ...	0	12	6
Live Cage	0	3	0
Stage Forceps	0	3	0
Camera Lucida ...	0	12	6
„ „ Wollaston's	1	4	0
Lieberkuhn			
Stand Condensers 5s. 9d., 7s. 6d., 15s.,	1	0	0
Turn-table ... from	0	4	0

	£	s.	d.
Section Cutting Machines from	0	9	0
Air Pump, with tray, &c. „	0	15	0
Microscopic Object Boxes 1s. 4d., 2s., 2s. 6d., 3s. 6d., 4s.,	0	9	0
„ Object Cabinets, mahogany, 40s.	3	0	0
„ Lamps, with Shades 6s., 10s. 6d., 14s.	1	0	0
„ Dissecting Knives, Scissors, Forceps, &c.			

All **Microscopic Accessories**, including objects (mounted and unmounted), supplied to order on the most reasonable terms.

TRADE DISCOUNT LIST ON APPLICATION.

'OPTIMUS' ANEROID BAROMETERS.

	s.	d.
Watch size, (1½ or 2 inch dial), Hard Enamel 26s. 6d.; Engraved Silvered	30	0
" with mountain scale, Enamelled 28s. 6d.; Engraved silvered 5,000 feet 33s.; 10,000 feet 36s.; 12,000 feet ...	42	0
" first quality compensated silvered dial	42	0
" first quality mountain scale silvered dial, 10,000ft. 50s.; 12,000ft. 54s.; 15,000ft. ...	65	0
Pocket size,(2½ or 3inch dial),compensated	48	0
Pocket size, with mountain scale, compensated, 10,000ft, 66s.; 15,000 feet, 75s.; 20,000 feet	83	0
Keyless motion for revolving altitude scale ... ... (extra)	8	0
Thermometers on dials ... ... "	3	0

Gold or Silver cases, hall-marked, extra according to weight.

Mining and Surveying Aneroid supplied to order.

ANEROIDS IN BRASS CIRCULAR MOUNTS and outside Boxes.

	5-inch.		Superior 6		8			Extra quality, with Thermometer 5		6		8	
	s.	d.	s.	d.	s.	d.		s.	d.	s.	d.	s.	d.
Card Dials ...	12	6	22	6	36	0	...	22	0	30	0	38	0
Enamelled Dials	22	0	29	0	44	0	...	32	0	42	0	54	0
Silv'rd engraved	22	0	33	0	49	0	...	32	0	42	0	58	0

	£	s.	d
Aneroids as above, fitted into carved oak and other wood cases, extra according to workmanship ... from 8s. 6d. to	5	0	0
Barograph by means of clock-work revolving drum carrying chart, the variations of atmospheric pressure are recorded ...	6	6	6

Kew verifications supplied with above instruments for a small extra charge.

MERCURIAL BAROMETERS.

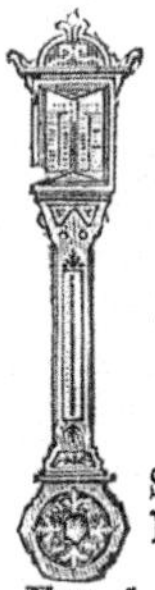

	£	s.	d.
Fitzroy, in wood frames ... 9s. 6d., 12s. 6d., 20s.	1	5	0
Ditto Carved ... £1 10s.; £1 15s.; £2 8s.	3	2	0
Wheel pattern frames 6in. dial 18s.; 8in. 26s.; 10in.	1	18	0
Ditto, Superior ... 30s.; 8in. 48s.; 10in.	3	8	0
Pediment, in Oak, Rosewood, Walnut or Mahogany, Fig. 1, £1 4s., £1 16s, £2 18s., £4.	7	10	0
Marine Barometer with Sympiesometer vernier and gymbals, Fig. 2 ... ... £3 15s.	4	10	0
Standard Barometer ... £7 10s., £12 12s.	22	10	0
Board of Trade	3	10	0

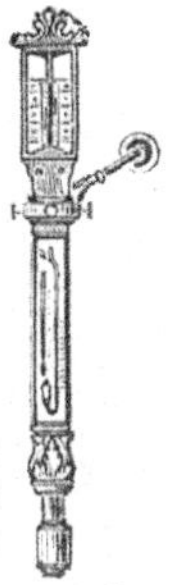

FIG. 1

FIG. 2

TRADE DISCOUNT LIST ON APPLICATION.

MATHEMATICAL INSTRUMENTS.

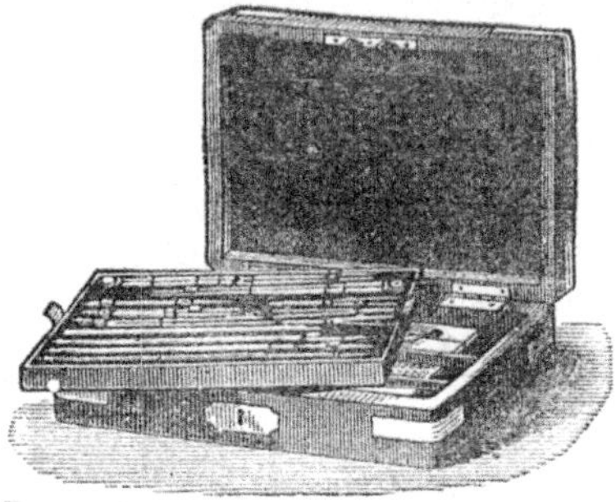

	£	s.	d.
Sets, French make, in wood boxes, complete, 1s., 1s. 3d., 1s. 9d., 3s., 6s., 10s.,	1	5	0
English, in Mahogany case, containing 6-in. electrum long-joint compass ink and pencil points and lengthening bar, ink and pencil bows, drawing pen, protractor, and ebony parallel ... each	1	10	0
Rosewood case, containing 6-inch electrum long-joint compass, ink and pencil points and lengthening bar, divider, ink and pencil bows, drawing pen, ivory protractor, and ebony parallel	2	2	0
Rosewood case, containing the following long-joint instruments:—6-inch compass with ink and pencil points and lengthening bar, divider, ink and pencil bows, set of three spring bows, two drawing pens, ivory protractor, and parallel	2	12	0

'OPTIMUS' THERMOMETERS.

		s.	d.		s.	d.
Clinical Indestructible, in boxwood or German silver cases	4 and 5 in.	3	0	6 in.	3	6
,, Indestructible Magnifying lens fronts	4 and 5 ,,	6	6	6 ,,	7	6

Chemical, for immersion, 150°, 1s.; 220°, 1s. 6d.; 300°, 3s.; 400°, 4s.; 600°, 5s

FIG. 1.

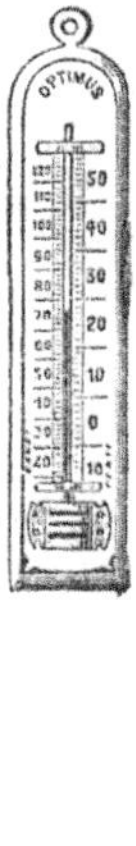

FIG. 2.

	8-inch s.	d.	10-inch s.	d.	12-inch s.	d.
Air, mounted on boxwood, spirit or mercury, enamelled tubes ...	0	5	1	0	2	0
,, superior, mounted on boxwood, spirit or mercury, enamelled tubes	1	6	2	6	3	6
,, Minimum, spirit, enamelled tubes	0	9	1	6		
,, Minimum, superior, spirit, enamelled tubes...	2	0	3	0		
,, Minimum, on porcelain, spirit	1	10	4	0	6	6
,, superior, on porcelain, spirit Fig. 1	5	0	6	6	8	0
,, Sixes, on boxwood	5	0	6	6	8	0
,, ,, on porcelain or opal glass	9	0	11	0	12	9
Bath, or Brewer's, on metal scales, Fig. 2	1	0	1	6	2	6
Ditto, on porcelain scales, Forbes specifications			3	3		
Ditto, metal scales, copper cases ...	1	10	3	0	5	6

OPINIONS OF 'OPTIMUS' PHOTOGRAPHIC LENSES.

We now turn to the "Optimus Rapid Euryscope," manufactured by the firm of Perken, Son, and Rayment, Hatton Garden, an example of which is on a camera on our editorial table. With its full aperture of 1¾ in. (its equivalent focus being 11 in.) it defines with extreme brilliancy, and when used with a stop it easily covers a 10 by 8 plate to the corners, which is larger than that engraved on the mount as its possibility. Working as it does with such a large aperture (*f* 6 approx) it serves as a portrait and group Lens, as well as a landscape and copying objective. There is no doubt of its proving a most useful lens, J. TRAILL TAYLOR.—*British Journal of Photography.*

Frith's series of "Life in London Streets" were all taken with 12 × 10 *Portable* "OPTIMUS" lens on Cobb's plates. Had an "Optimus" *rapid* symmetrical been used, the plates would have been over-exposed with full aperture.—G. LINDSAY JOHNSON, M.A.,M.B., F.R.C.S., England.—*Amateur Photographer.*

PORTRAIT LENS.—The "Optimus" lenses are moderate in price, and yield most excellent results.—*Amateur Photographer.*

"OPTIMUS" LENS.—I have taken trains going at 30 miles an hour, and think the lenses perfect for rapidity and definition.—E. J. WALL.—*Amateur Photographer.*

. . . . We may call attention to the extensive optical and metal works that Perken, Son, and Rayment have established in Hatton Garden, and their photographic cabinet factory in Saffron Hill. At the former we were much interested in the glass-grinding departments—one for photographic lenses, another for spectacles: and we were surprised to find in London such extensive workshops for the metal parts of cameras and optical lanterns; indeed, we thought outside Birmingham we should not find such workshops in the United Kingdom. The cabinet works in Saffron Hill also interested us much; the arrangement of the machine tools, and distribution of power on the several floors, being admirable.—*Photographic News.*

We are pleased to find upon trial that the Lens ["Optimus" Rapid Rectilinear] sent for review is really an excellent instrument.—*Photographic News.*

If E. G. E. wants the finest lens in the market he cannot do better than get the "OPTIMUS."—Professor DE FRERE.—*Amateur Photographer.*

"OPTIMUS EURYSCOPE."—I am much pleased with the 9 by 7 received last week. It covers 12 by 10, and gives splendid definition. For a good, useful, all-round lens I consider it perfect.—E. BRIGHTMAN, Hon. Sec., Bristol and West of England A. P. A.

"OPTIMUS EURYSCOPE."—When in your establishment you kindly lent me one of your 7 by 5 I like the lens so well, I do not want to part with it.—CHAPMAN JONES, F.I.C., F.C.S.

"OPTIMUS EURYSCOPE."—This lens is of extra large diameter; the aperture being *f* 6, and while admitting of the most rapid exposures, gives fine marginal definition. We have obtained with it excellent portraits in an ordinary sitting room. The lens is a most useful one all-round.—T. C. HEPWORTH.—*The Camera,* April 1, 1889.

As I have often said before, I consider these lenses unequalled.—E. J. WALL, *Amateur Photographer.*

"I have tried the 7 × 5 'OPTIMUS' Euryscope. It is a very satisfactory lens and covers the *whole plate* for interiors with *f* 16; being able to work at *f*/6 makes it very handy for portraits."—J. G. P. VEREKER.

IRIS DIAPHRAGM Fitted to Rapid Euryscope and „ Rectilinear	5 by 4	7 by 5	9 by 7	10 by 8	12 by 10	15 by 12	18 by 16
	10/-	11/-	12/6	14/-	16/-	18/-	20/-

Extra.

'OPTIMUS'

RAPID PHOTOGRAPHIC LENSES.

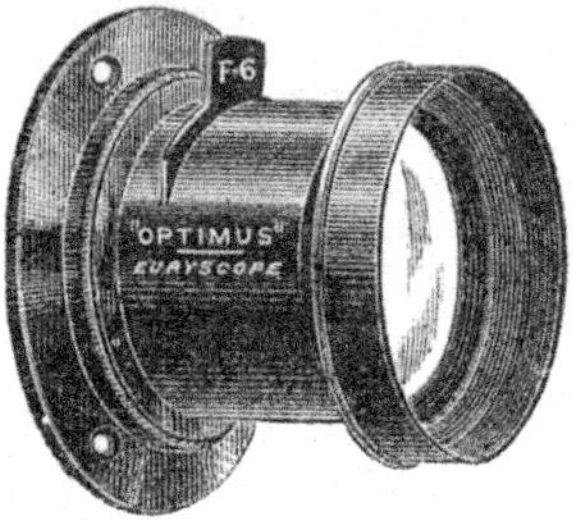

Extra Rapid Euryscope, large Diameter (Double). The aperture is **F 6.** The Lenses are of special optical glass, constructed with the nicest precision of curvatures, so maintaining good marginal definition, coupled with the most **Extreme Rapidity.**

To cover	5 by 4	6 by 5	7 by 5	8 by 5	9 by 7	10 by 8	12 by 10
Equiv. Focus	5¼	6½	8¾	10	12	14	18 inches.
	63/-	**78/-**	**94/6**	**110/-**	**126/-**	**220/-**	**380/-**

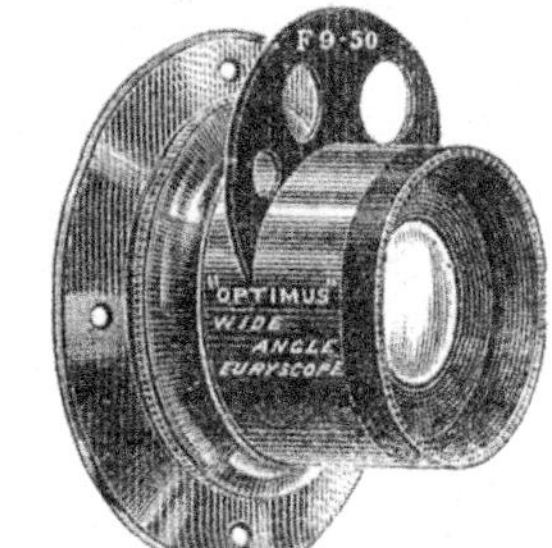

Wide Angle Euryscope (Double), F 9·50. This aperture is exceedingly open for wide angle work. The definition, however, is in no way sacrificed, as the curvatures are perfectly accurate, and the most minute detail in architectural and interior subjects is rendered with the maximum of crispness, and a total absence of distortion.

To cover	5 by 4	7 by 5	9 by 7	10 by 8	12 by 10
Equiv. Focus	4	4¾	6	8	10 inches.
	63/-	**94/6**	**126/-**	**220/-**	**380/-**

TRADE DISCOUNT LIST ON APPLICATION.

PERKEN, SON, & RAYMENT, 99, Hatton Garden, LONDON

IRIS DIAPHRAGM Fitted to Rapid Euryscope and ,, Rectilinear	5 by 4	7 by 5	9 by 7	10 by 8	12 by 10	15 by 12	18 by 16
	10/-	**11/-**	**12/6**	**14/-**	**16/-**	**18/-**	**20/-**

Extra.

Portable Symmetrical (Double), F 16 with revolving diaphragms. Specially adapted for Architecture, being of short focus and wide angular aperture, can be used to advantage when very close to the subject. It is also useful for landscapes, as well as copying. The smaller sizes give beautiful **Lantern Slides,** the definition being exceptionally crisp.

To cover	$\frac{5}{4}$	$\frac{7}{5}$	$\frac{9}{7}$	$\frac{10}{8}$	$\frac{12}{10}$	$\frac{15}{12}$	$\frac{18}{16}$	plates.
	39/-	52/6	82/6	127/6	142/6	180/-	225/-	

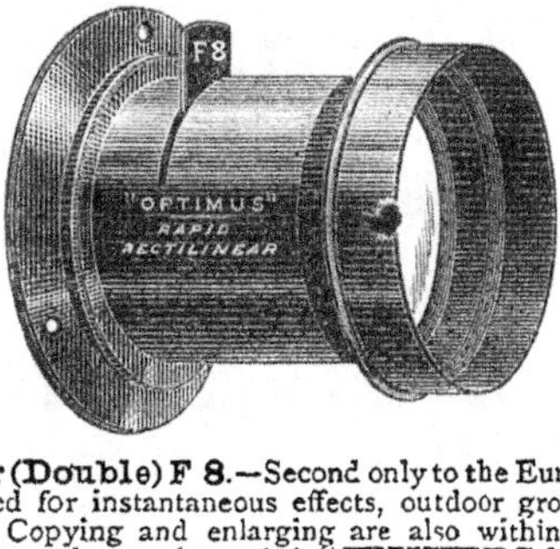

Rapid Rectilinear (Double) F 8.—Second only to the Euryscope for Rapidity therefore well suited for instantaneous effects, outdoor groups and views, as well as interiors. Copying and enlarging are also within the capabilities of the lens; in fact its work may be styled "**UNIVERSAL.**"

To Cover	$\frac{5}{4}$	$\frac{6}{5}$	$\frac{7}{5}$	$\frac{8}{5}$	$\frac{9}{7}$	$\frac{10}{8}$	$\frac{12}{10}$	$\frac{15}{12}$	$\frac{18}{16}$	
Focus	$5\frac{1}{2}$	$6\frac{3}{4}$	$8\frac{3}{4}$	10	12	14	18	20	25	inches
	33/-	**45/-**	**49/6**	**64/-**	**82/6**	**127/6**	**142/6**	**180/-**	**225/-**	

. . . We may call attention to the extensive optical and metal works that Perken, Son and Rayment have established in Hatton Garden, and their photographic cabinet factory in Saffron Hill. At the former we were much interested in the glass-grinding departments—one for photographic lenses, another for spectacles; and we were surprised to find in London such extensive workshops for the metal parts of cameras and optical lanterns; indeed, we thought outside Birmingham we should not find such workshops in the United Kingdom.—*Photographic News.*

We are pleased to find upon trial that the Lens ["Optimus" Rapid Rectilinear] sent for review is really an excellent instrument.—*Photographic News.*

TRADE DISCOUNT LIST ON APPLICATION.

Rapid Landscape.—Works **F 11**, and gives brilliant negatives. Particularly suited for landscapes also capable of being worked as a Portrait lens.

To cover...	5 by 4	7 by 5	9 by 7	10 by 8	12 by 10 plates.
Focus ..	5½	9	12	14	18 inches.
	25 -	**36 -**	**45/-**	**75 -**	**110/-**

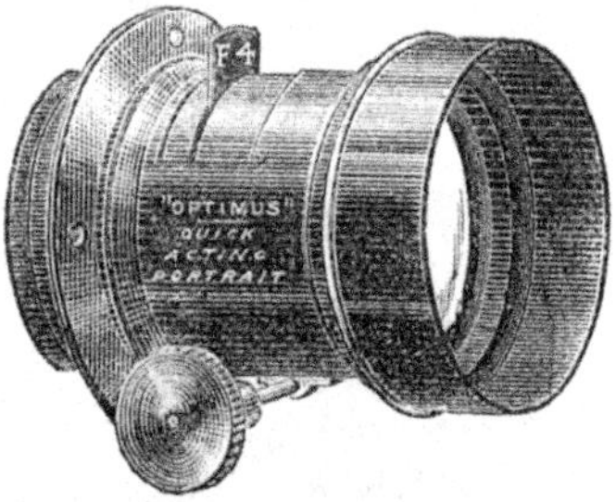

Portrait Lens.—Specially constructed as quick acting for short exposures in Portraiture. They are second to none, the definition being maintained by their perfect optical qualities.

Diam.	2 inches.	2¾ inches.	3¼ inches.	
Price	90 -	120 -	180 -	*Larger Sizes to order.*
	1 B.	**2 B.**	**3 B.**	

'DEAR SIRS.—Herewith your 7 x 5 "Optimus' Lens, which, as per your request, I have tried in the production of large heads. Along with it I send two negatives taken by it, the head in one of them measuring *two* inches, that in the other being *three* inches. In both, the perspective seems right enough, there being no appearance of its being strained or violent.

'With us, it was rather dark and very rainy all day, and I took the negatives inside a room, without a diaphragm, exposure 7 secs. and 10 secs. respectively. The SHARPNESS of all the planes of the head is good, as you will perceive. The distance of the sitter from the lens was 3 ft. 6½ ins. for the larger head, and 4 feet 9 in. for the smaller head.—Yours truly, J. TRAILL TAYLOR.'

'If E. G. E. wants the FINEST LENS in the market he cannot do better than get the "OPTIMUS."—Professor DE FRERE.'—*Amateur Photographer.*

TRADE DISCOUNT LIST ON APPLICATION.

PERKEN, SON, & RAYMENT, 99, Hatton Garden, LONDON.

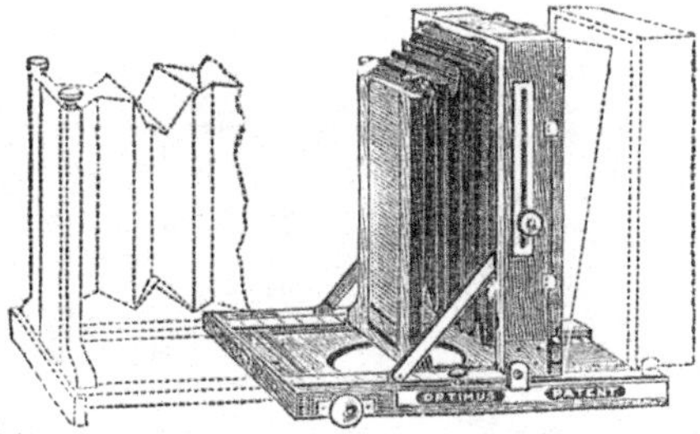

"OPTIMUS" WIDE ANGLE CAMERA. EXTRA LONG FOCUS.

Baseboard does not cut off the view when using wide angle lenses

This Instrument possesses every possible advantage, being very Light, very Rigid, and very Portable. The focussing screen and body may be brought towards the front of Baseboard so as to prevent obstruction when using lenses of wide angular aperture and short focus. It is provided with double-swing arrangement and long-focussing (rack) adjustment. When closed the lens may remain attached to its proper position (the front), and project through the TURNTABLE Baseboard.

Price, including 3 Double Dark Slides—

4¼ by 3¼	5 by 4	6½ by 4¾	8½ by 6½	10 by 8	12 by 10	15 by 12
140 -	**146/-**	**165,-**	**188 -**	**235 -**	**288/-**	**350,-**

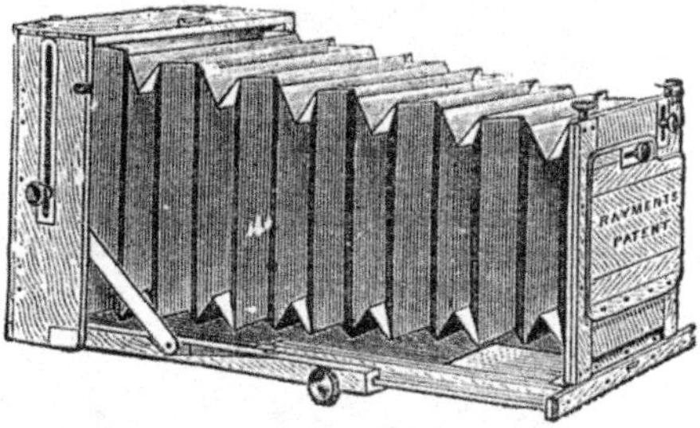

RAYMENT'S PATENT CAMERA. EXTRA LONG FOCUS

"I should strongly recommend RAYMENT's Camera. It is LIGHT, COMPACT, very RIGID, and extends to about double the usual focus."—*Amateur Photographer.*

"The 'RAYMENT' Camera, in particular, claims attention both for its BEAUTY of WORKMANSHIP and for the EASE and READINESS with which it can be put into action."—*The Camera.*

It can be set up a most instantaneously, has no loose parts, and includes all motions, having hinged focussing screen (adjusted by rack and pinion action), double swing back, cross fronts reversing back arrangement so that oblong dark slides give either horizontal or vertical pictures without unscrewing the Camera from the tripod. Price of Camera, including 3 Double Dark Slides—

4¼ by 3¼	5 by 4	6½ by 4¾	8½ by 6½	10 by 8	12 by 10	15 by 12
120 -	**126 -**	**145 -**	**168 -**	**212 -**	**258 -**	**314/-**

TRADE DISCOUNT LIST ON APPLICATION.

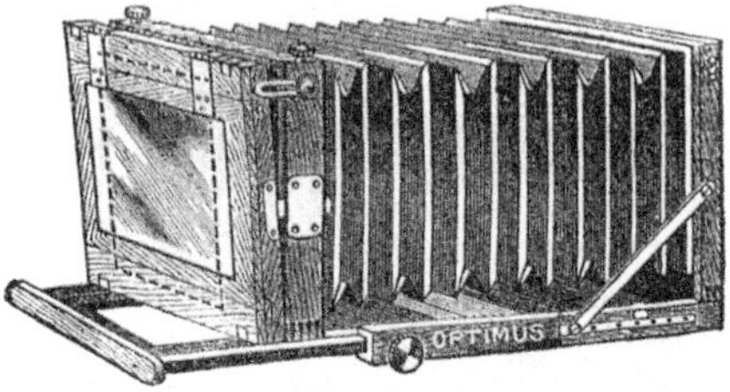

"OPTIMUS" CAMERA. LONG FOCUS

The Instrument can be set up almost instantaneously, has no loose parts, and includes all motions, having hinged focussing screen (adjusted by rack and pinion action), double swing back, cross fronts reversing back arrangement so that oblong dark slides give either horizontal or vertical pictures without unscrewing the Camera from the tripod.

Price of Camera, including 3 Double Dark Slides –

4¼ by 3¼	5 by 4	6½ by 4¾	8½ by 6½	10 by 8	12 by 10	15 by 12
130/-	133 -	137/-	175/-	227/-	275/-	333/-

"OPTIMUS" STUDIO CAMERA.

Specially arranged for Studio Use.

"Invited to say, if in our estimation, the Studio Cameras of Perken, Son, and Rayment could be improved in any way whatever, for the purpose for which they are intended, we must answer, No!" *British Journal of Photography.*

This Camera is perfectly rigid, has double length of Bellows to suit small studios where large pictures are required, is fitted with Mechanical Adjustment to Focus, and swing back. It is in all respects a perfect Instrument.

Camera with repeating frame masks, and one single dark slide—

6½ by 6½	8½ by 8½	10 by 10	12 by 12	15 by 15
145/-	188 -	225 -	265/-	325/

TRADE DISCOUNT LIST ON APPLICATION.

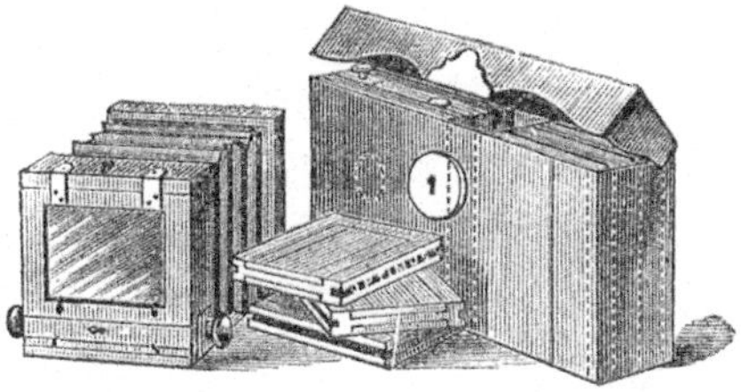

"OPTIMUS" PORTABLE FOLDING CAMERA.

Cheap, Strong, Serviceable, and Efficient.

Durable bellows, hinged focussing screen with sliding adjustment, readily and securely held in exact position by means of a pinion passing through the body nearest bottom or baseboard and having milled head screws on either side of the instrument to clamp tight.

Camera and one Double Dark Slide—

4¼ by 3¼	5 by 4	6¼ by 4¾	8½ by 6½
21/-	**24/-**	**39/6**	**48/-**

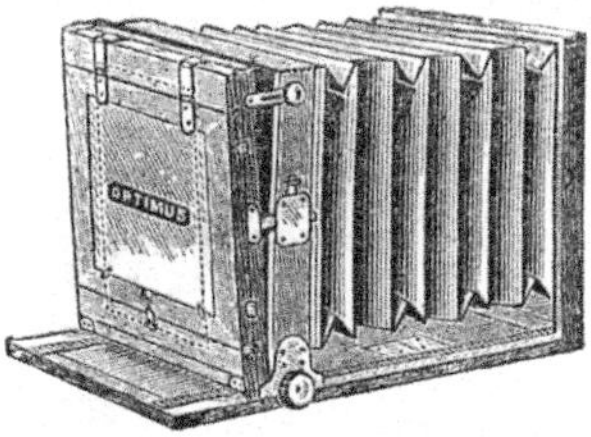

PORTABLE (A.R.) CAMERA.

Compact, Rigid, Inexpensive, and of Excellent Finish'

These Instruments have Leather Bellows, and are fitted with hinged Focussing Screen, adjusted by Rack and Pinion, Square Reversing Back, so that horizontal or vertical pictures may be taken without removing Camera from Stand.

Price of Camera, including 3 Double Dark Slides—

4¼ by 3¼	5 by 4	6¼ by 4¾	8½ by 6½	10 by 8	12 by 10	15 by 12
85/-	**87/6**	**98/-**	**125/-**	**150/-**	**200/-**	**260/-**

Extra for Brass Binding Camera, and 3 Double Dark Slides—

27/6	**28/-**	**30/-**	**32/-**	**33/-**	**38/**	**48/**

TRADE DISCOUNT LIST ON APPLICATION.

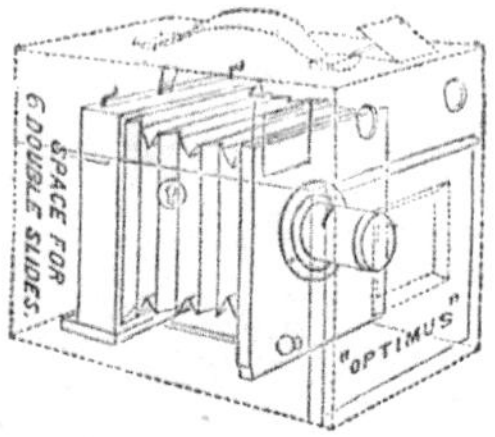

"OPTIMUS" DETECTIVE CAMERA.

Price, including three Double Dark Slides.

	£	s.	
With Optimus Rapid View Lens Working F8	5	10	0
Ditto, with "Optimus" Rapid Rectilinear working F8	6	6	0
Ditto, with "Optimus" Rapid Euryscope working F6	7	7	0
"Extra" Double Dark Slides each		8	6

The Shutter is arranged for exposures of any duration not less than $\frac{1}{100}$ of a second. The Shutter and Camera occupy so little space that six Double Dark Slides, accommodating twelve Dry Plates, can be carried, in addition to a Focussing Screen. A panel slides up at the end of the box, displaying the screen for focussing. The exactness in focussing is simple, and the position is maintained by a secure clamping arrangement. A similar sliding panel is fitted to the front end, which completely hides the lens. The exterior is covered with leather, and measures $9\frac{1}{4}$ by $5\frac{1}{2}$ by $7\frac{1}{2}$. (Negatives $4\frac{1}{4}$ by $3\frac{1}{4}$ or $3\frac{1}{4}$ by $3\frac{1}{4}$.

Carries twelve dry plates.

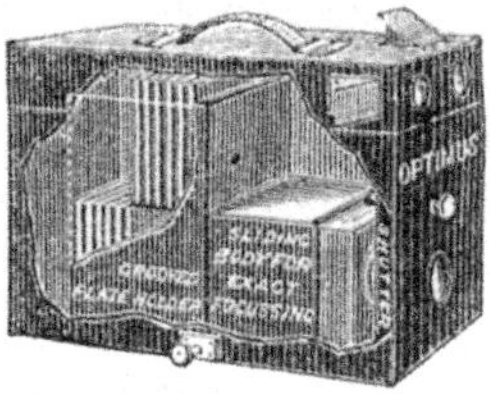

Plates changed by turning a button.

"OPTIMUS" MAGAZINE CAMERA.

NO DARK SLIDES REQUIRED.

With Rapid Rectilinear Lens, working at F8	£6	6	0
With Rapid Euryscope working at F6	7	7	0

Twelve dry plates are placed in the upper portion of the grooved travelling reservoir. The bases of these plates rest upon the top of the Camera body. The grooved reservoir recedes gradually from the Exposure Chamber when the pinion is revolved—allowing one plate at a time to fall to the bottom half of the reservoir, and so place its sensitized surface within the Exposure Chamber opposite to the Lens, and exactly in true focussing register. The rapidity and simplicity of working is unique. The twelve plates may, if desired, be exposed in as many seconds. There is no possibility of the plates sticking. The Shutter is suitable and convenient.

TRADE DISCOUNT LIST ON APPLICATION.

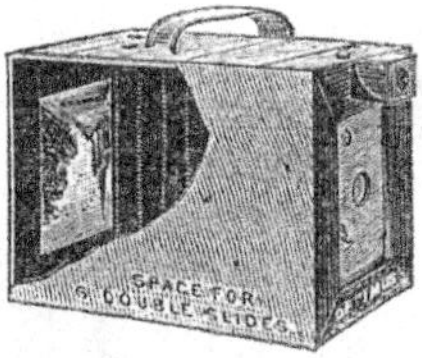

"OPTIMUS" UBIQUE HAND CAMERA.

FOR PLATES 4¼ by 3¼

OUTSIDE DIMENSIONS OF CAMERA, 8 by 4½ by 6 INCHES.

This instrument has 3 Double Dark Slides and is self contained. The shutter gives rapid or slow exposures. With View Finder, Focussing Screen, and

"Optimus" Rapid View Lens, Complete 55/-

"Optimus" Rapid Rectilinear Lens, Complete 80/-

"OPTIMUS" MINIMUS HAND CAMERA (PATENT)

FOR PLATES 4¼ by 3¼

OUTSIDE DIMENSIONS OF CAMERA 8 by 3½ by 5 INCHES.

This Camera has no Dark Slides, carries Twelve Dry Plates. They place themselves consecutively before the Lens; each, after exposure, is removed to the back of the others by means of the little bag underneath the instrument. The register indicating the number of plate offered for exposure, renders it impossible for the same plate to be twice exposed. It is provided with Focussing Adjustment and View Finder, also with a Curtain Shutter which gives slow or rapid exposures.

Price with "Optimus" Rapid Euryscope £7 7 0

TRADE DISCOUNT LIST ON APPLICATION.

PERKEN, SON, & RAYMENT, 99, Hatton Garden, LONDON

"OPTIMUS" (ECONOMIC, PORTABLE,) CAMERA (LONG FOCUS.)

STRONG, SERVICEABLE AND EFFICIENT.

It has long extension Taper Bellows, Rising Front, Rack Adjustment, Square Reversing Frame, and other modern improvements.

Price with Three Double Dark Slides—

4¼ by 3¼	6½ by 4¾	8½ by 6½
93 -	**114/-**	**136 -**

"OPTIMUS"

ECONOMIC PHOTOGRAPHIC SETS.

The above Camera is supplied wlth Dark Slide, Rapid Rectil'near Lens Instantaneous Shutter and Tripod, complete with waterproof case.

4¼ by 3¼	6½ by 4¾	8½ by 6½
84/-	**110 -**	**160 -**

TRADE DISCOUNT LIST ON APPLICATION.

"OPTIMUS"

PORTABLE SETS OF PHOTOGRAPHIC APPARATUS.

Fitted COMPLETE in Cabinet, with Tripod Stand.

	£	s.	d.
Quarter Plate Size (4¼ X 3¼) includes every requisite	1	16	0
Half Plate Size (6½ X 4¾)	3	10	0
Quarter Plate Size (4¼ X 3¼) as diagram	2	5	0
Half Plate Size (6 X 4¾)	5	0	0
Superior Camera, Rack and Pinion focussing adjustment,			
Quarter Plate (4¼ X 3¼)	3	8	0
,, Half Plate (6½ X 4¾)	5	15	0
,, Whole Plate (8½ X 6½)	9	10	0

BRASS BINDING *any* CAMERA, and 3 Double Dark Slides--

4¼ by 3¼	5 by 4	6½ by 4¾	8½ by 6½	10 by 8	12 by 10	15 by 13
27 6	28 0	30 0	32 0	33 0	38 0	48 0

EXTRA DOUBLE DARK SLIDES for *any* of our Cameras--

	4¼ by 3¼	5 by 4	6½ by 4¾	8½ by 6½	10 by 8	12 by 10	15 by 12
Solid, no hinges— EACH	7 3	8 3	10 9	...	...	...	...
Solid, with hinged shutters EACH	10 3	10/9	11/9	...	...	...	..
Hinged opening and Hinged shutter EACH	14 -	14 6	15/-	21 6	25 6	31/6	47/3

TRADE DISCOUNT LIST ON APPLICATION.

CAMERA CASES, with Shoulder Straps,

LINED WITH GREEN CLOTH.

SQUARE— $\frac{1}{2}$, 20/-; $\frac{1}{1}$, 29/-; $\frac{10}{8}$, 35/-; $\frac{12}{1}$, 46/- Leather.
$\frac{1}{2}$, 15/9; $\frac{1}{1}$, 20/-; $\frac{10}{8}$, 21/6; $\frac{12}{1}$, 28/- Canvas.

OBLONG— $\frac{1}{4}$, 10/6; $\frac{1}{2}$, 15/-; $\frac{1}{1}$, 20/- Canvas only.

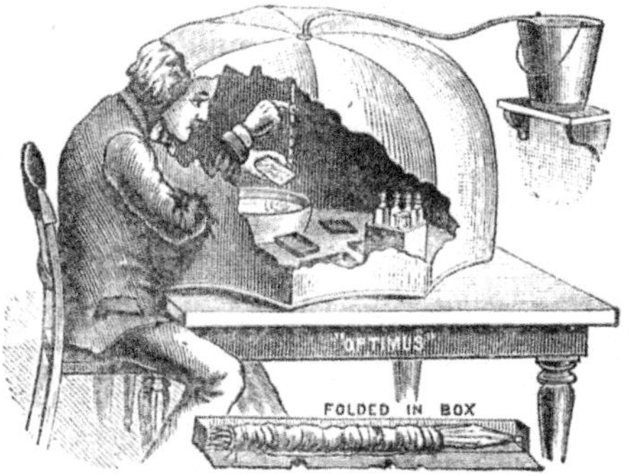

PORTABLE "UMBRELLA" RUBY TENT (PATENT NON-ACTINIC.

Like an umbrella, it folds into a very small space, and can be set up for developing or changing plates as easily as its well-known prototype. Made of two thicknesses of material, one ruby, the other orange colour, no light can enter except it be filtered through these media. The head and hands are introduced so that the operator, sitting in his chair can conveniently watch the progress of his work whilst the tent rests on the table. *Beginner's Guide to Photography.*

Made in two sizes—	Dimensions closed. Inches	Price s. d.
For changing Plates	24 by 3	25 0
For Developing ,, ...	28½ by 3	35 0

TRADE DISCOUNT LIST ON APPLICATION.

WAISTCOAT DETECTIVE CAMERA.

THIS INSTRUMENT IS ONLY MADE BY THE PATENTEE,
C. P. STIRN.

With Plates for 36 Exposures	... ..	23 6
" 24 " Larger Size for Lantern Pictures	..	35 -

"OPTIMUS" VIEW FINDERS.

Camera Obscura Model	...	... each 5/-
1¼-inch bi-concave	...	... " 2 6
2¼ " "	...	... " 6 -
3¼ " "	...	... " 8 6

FOCUSSING GLASSES.

Per doz.—

9 9 15 9 18 3 33 9

With Archimedean Screw.

60 - per doz.

COMBINED FOCUSSING GLASS AND FINDER.

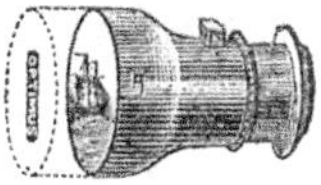

Bell Shape,
5/- each.

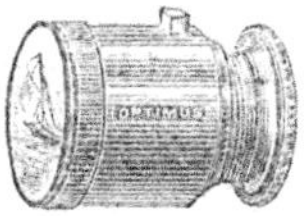

Screw Adjustment.
4/- each.

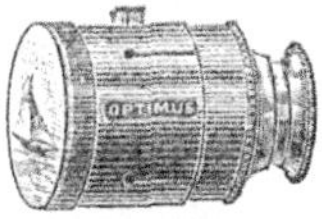

Sliding Adjustment
5/- each.

TRADE DISCOUNT LIST ON APPLICATION.

PATENT

PERFECTION

PHOTOMETER.

CORRECT EXPOSURE A CERTAINTY.

Each ... **6 9.**

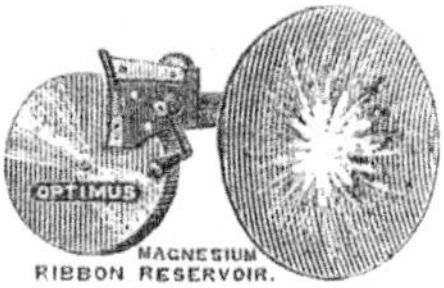

"OPTIMUS"

MAGNESIUM

RIBBON LAMP, 5/-

British Journal of Photography, Nov. 7th, 1888 :—" It is a neat little thing, not greatly exceeding the dimensions of an old-fashioned watch, and projects a *powerful* beam of *light*.

HASTINGS'

FLASH LAMP.

1/3 With Mouthpiece. | With Pneumatic Ball **2 6**

" A handy Lamp. It is especially adapted for taking instantaneous photographs at night. The glass tube is charged with magnesium powder, and the brass trough filled with methylated spirit; this being lighted, the pneumatic ball is squeezed, the flash is given, and the photograph taken."—*Amateur Photographer.*

MAHOGANY

RETOUCHING DESKS.

Whole-plate	16 8
12 x 10	25 -
15 x 12	35/6

Each has the smaller Carriers.

TRADE DISCOUNT LIST ON APPLICATION.

"OPTIMUS" PLUNGE SHUTTER (PATENT).

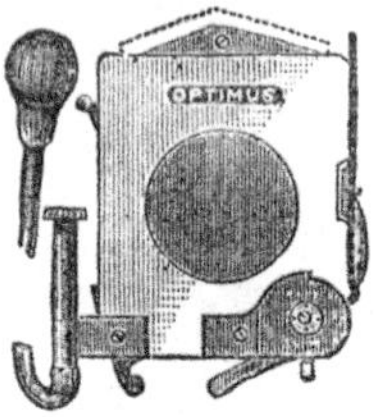

"Mr. J. Traill Taylor exhibited a pneumatic shutter, giving exposures of any duration, at will, and av'iding all vibration. He stated it was one of the best shutters he had seen, and said it was made by Perken, Son, and Rayment.'—*British Journal of Photography.*

This may be styled the most **PORTABLE** of shutteis. It is made either to fit on the hood, or may be adapted to act between the lenses of a doublet.

To fit ¼ or ½ plate Optimus Lens	30	Complete with Pneumatic Release.
$\frac{1}{1}$ " " "	36	
" 10 × 8 " "	48	

"OPTIMUS" BETWEEN LENS SHUTTERS.

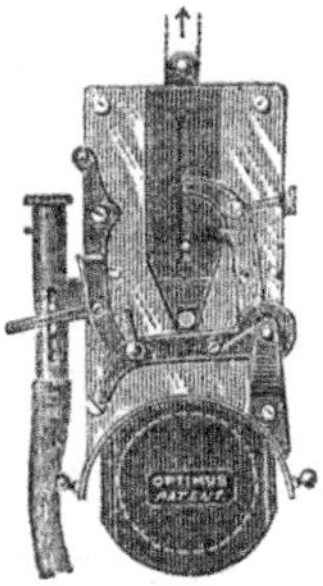

Is arranged to fit the lens mount like a saddle. Exposure is effected by a plunger working between the two combinations of a double lens, or in front of them if preferred. Both pneumatic and hand releases are provided. Exposures varying from $\frac{1}{75}$ of a second to prolonged may be attained. The working parts are strong, and derangement impossible with a reasonably careful operator. Weight and bulk are reduced to a minimum.

Prices same as above.

PHANTOM SHUTTER, with Hand Release.

For Diameter of Front Lens.

Inch.	1½	1¾	2	2¼	2½	2¾	3	3¼	3½
	13/-	14 6	15 9	16 9	18 3	19 6	20 9	21 9	23 3

IF FITTED WITH PNEUMATIC RELEASE.

	1½	1¾	2	2¼	2½	2¾	3	3¼	3½
	20 9	22 -	23,3	24 6	25 9	27 -	28 3	29/6	30/9

TRADE DISCOUNT LIST ON APPLICATION.

TRIPOD

AND

STUDIO.

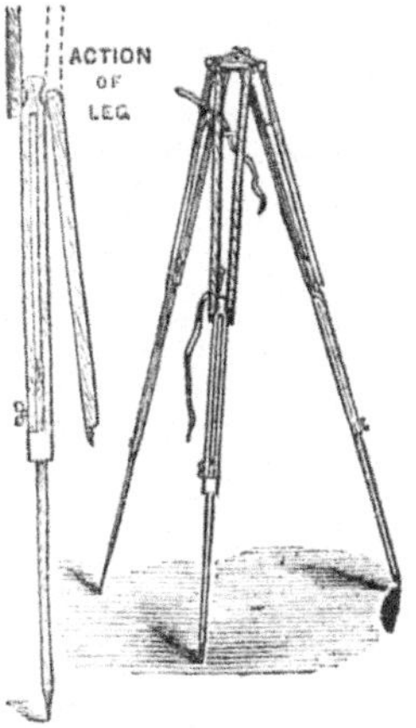

"OPTIMUS" STANDS FOR CAMERAS.

	1/4 s.	d.	1/2 s.	d.	1/1 s.	d.	10/8 s.	d.
Telescopic, with sliding leg adjustment, rigid	8	6	9	6	10	9	16	9
Folding Ash, with Bayonet joint, "E.P., very rigid	9	9	10	9	12	6	20	3
Telescopic Ash, with sliding leg adjustment, "Maudsley" pattern	18	6	19	9	20	9	24	9
Telescope Ash, 3-fold, as sketch			22	0	24	0	29	0
'OPTIMUS' Stand, very rigid and much recommended for large sizes			15	0	16	0	22	6
*4-FOLD, Very Portable			20	0	25	0	35	0
Pine Studio Stands with clamp	14	6	30	0	RACK		43	9
Mahogany					45	0	78	9

** Specially Portable. Strongly recommended where Small Bulk is important.*

"OPTIMUS"

MICRO-PHOTOGRAPHIC APPARATUS.

	£	s	d
Camera, with Dark Slide, Micr scope, 2-inch Objective, Lamp, Condenser, &c.	9	10	0
Superior Camera and Microscope, with Rack and Pinion Adjustment Fitted for Focussing as well as for Adjusting the Stage	14	0	0

HEAD RESTS.

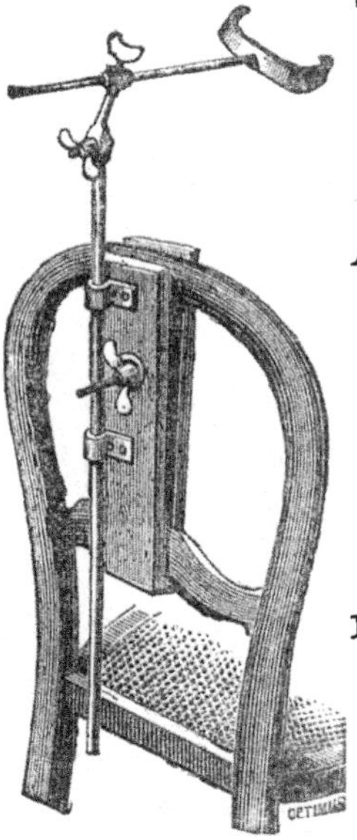

Amateur, to fit on Chair ... 7 6

STUDIO

15/- 28- 42/-

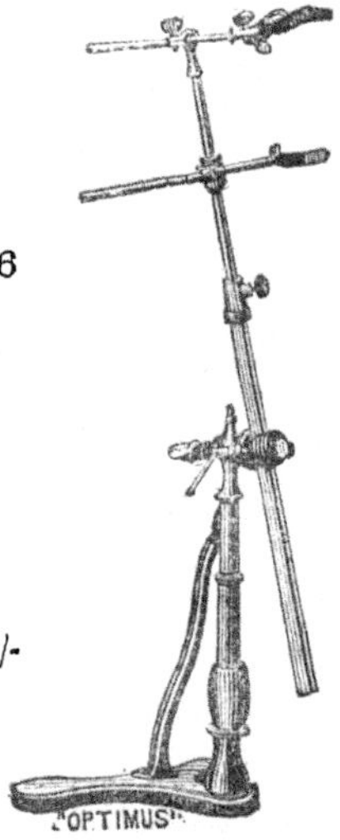

TRADE DISCOUNT LIST ON APPLICATION.

OPTIMUS INFALLIBLE DEVELOPER.

THE Components of this Developer are perfectly pure, and the exact quantities necessary are employed, so ensuring the best possible negatives, whether Landscapes or Portraits, be the Exposures Instantaneous or Prolonged.

TO PREPARE THE DEVELOPER.

Place the contents of the three packets into a quart bottle and fill up with 40 ozs. pure (if possible distilled) water. When all is completely dissolved it is ready for use.

TO ECONOMISE THE DEVELOPER.

After developing, pour the used Developer into a second bottle, for it can be advantageously employed in developing other plates.

To obtain the best possible results we recommend the following instructions to the careful attention of the operator:

For Plates—

Over Exposed	... ...	**use Old Developer only.**
Instantaneously Exposed		,, **New Developer** ,,
Under Exposed	... ...	,, **half old & half new.**

Place the plate in the developing dish or tray and pour over it sufficient Developer to cover the upper surface thoroughly.

When the detail does not appear as quickly as expected add new Developer, or use new only, as may be found necessary.

IN PACKETS SUFFICIENT FOR 100 ¼-PLATES.

PRICE **1s. 6d.**

TRADE DISCOUNT LIST ON APPLICATION.

DAYLIGHT DEVELOPING DISH.

PATENT.

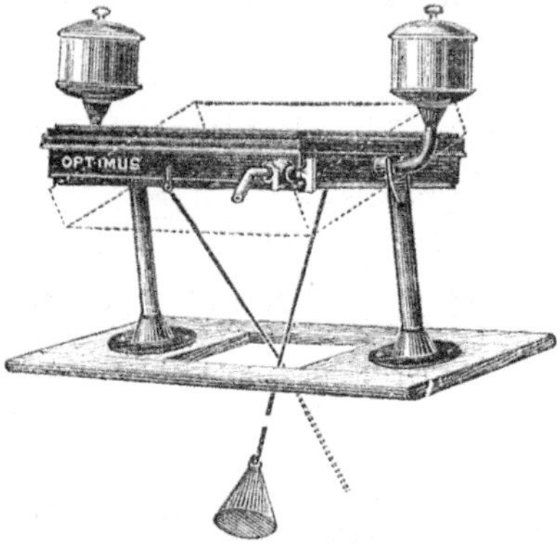

PERKEN, SON & RAYMENT, 99, Hatton Garden, LONDON.

WE would draw the attention of those who find the atmosphere and darkness of the dark room objectionable, to this apparatus which we have just completed, and which greatly reduces the time necessary for the operator to be imprisoned in the black hole—in fact, if a changing box or bag be used, the dark room is unnecessary.

It consists of a dish with non-actinic glass top and bottom, supported on a convenient stand, and having an AUTOMATIC ROCKING ARRANGEMENT, which keeps the dish oscillating during development

At the sides of the dish external funnels communicate with the interior, and are so arranged that white light cannot reach the plate. Through these funnels solutions are conveyed to the plate within, which may be strengthened or weakened at will, or run off by means of a waste pipe, and the plate flooded with water.

The fixing, which does not require the continued presence of the operator, may be accomplished in the changing bag or dark room.

Price for	...	**7 × 5**	...	...	**21/-**
"	..	**9 × 7**	...	...	**27/6**

TRADE DISCOUNT LIST ON APPLICATION.

"OPTIMUS" LAMPS FOR DARK ROOM.

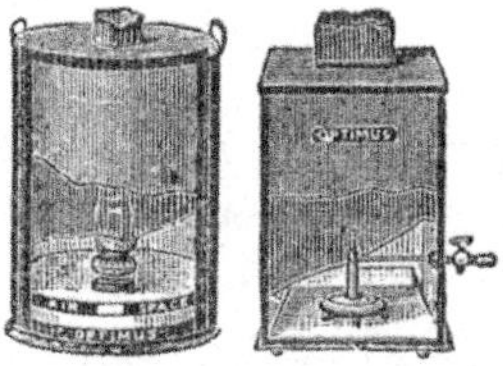

The Lantern shown in the diagram is fitted with a gas jet adjustable from the outside: the light can thereby be readily lessened or increased at will. In front is a sheet of ruby or orange glass, easily removed, behind which is a double thickness of canary fabric set in a metal frame; it is, therefore, safe when developing the most sensitive of plates. As development progresses, one of the non-acting media can be moved, and the negative examined by the protection the second medium continues to give, enabling the amount of detail to be judged with certainty. In this lamp the joints are all perfectly light-tight, being made with a double turn over of tin; the upper parts are also held together with rivets. Ventilation is well considered, as a shaft at the back of the lamp, open at bottom and top, encourages a free circulation of air. Without doubt this lamp has no equal for the purpose for which it is intended.—*British Journal of Photography.*

Fitted either with Gas or Argand Burner for Paraffin Oil.

Square or Round . . .. each **13s.**

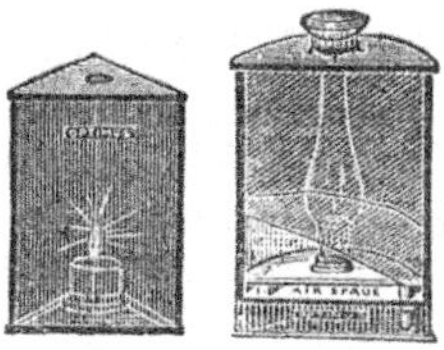

The form of Lantern shown in the accompanying diagram presents a great many advantages. It possesses a powerful Lamp so arranged that the oil receptacle is isolated from the flame and cannot get heated. Plenty of air circulates. In front is a sheet of ruby glass (removable), behind which is a sheet of deep orange, it is therefore safe when developing the most sensitive of Plates. As development progresses the ruby glass can be raised and the Negative examined by the orange glass only, enabling the amount of detail to be readily judged, per doz. **88/-**

FOLDING LAMP.—An inexpensive form of Lamp having two sides of metal, and the third of red glass. The metal sides are hinged together, so that they fold up for travelling, with the ruby glass protected from fracture by lying between them. Top and bottom triangular pieces—one forming a candle-holder and the other a chimney—complete this clever little arrangement ... ... per doz. **30/-**

REDDINGS' PATENT PORTABLE LAMP—
Small, **40/-**: Medium, **54/-**; and Large, **72/-** per doz.

CANDLES for above .. ... Small, **28/-**; Large ditto, **42/-** gross.

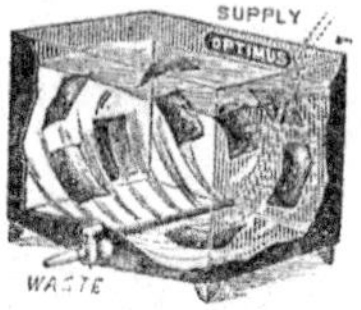

"OPTIMUS" REVOLVING PRINT-WASHER.

Water is injected from a perforated tube which crosses the tank at the bottom. The force of water creates a revolving current which carries the prints over and over in its course. The bottom is slightly V shaped and contains an outlet for waste.

Prices, including Grooved Metal Rack, which fits into the tank and accommodates negatives which can be also washed. Making the machine efficient for both NEGATIVES and PRINTS.

For Negatives and Smaller ... $\frac{1}{1}$, **16-**; $\frac{10}{8}$, **21-**; $\frac{12}{10}$, **30-**

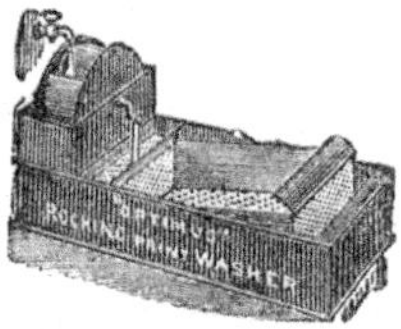

"OPTIMUS" ROCKING PRINT-WASHER.

Water running from a tap revolves the wheel which is connected with the cradle causing it to rock up and down at each revolution. A syphon is fitted to drain the tank. One hour's washing is ample.

inches 9 × 7 cradle	**18 6**	11 × 9 cradle	**28-**	13 × 11 cradle	**32/-**
16 × 13 ,,	**39 6**	20 × 16 ,,	**50-**		

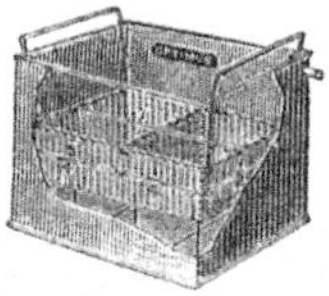

NEGATIVE WASHER

AND DRAINING RACK, COMBINED,

For Quarter, Five by Four, Half, and Whole Plate Negatives, **7/9.**

TRADE DISCOUNT LIST ON APPLICATION.

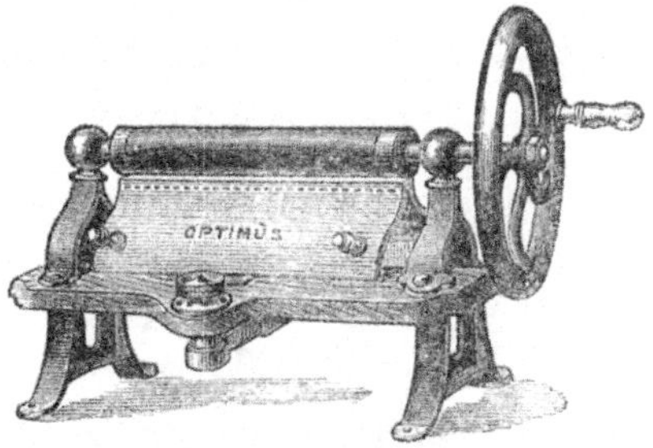

"OPTIMUS"
BURNISHERS FOR PHOTOGRAPHIC PRINTS

The Burnishing Bar is specially hardened. The Frame is of superior and convenient construction. It is Nickel-plated, and of beautiful finish. Each instrument is mounted on a table with legs.

Plate	$\frac{1}{4}$	$\frac{1}{2}$	$\frac{1}{1}$	$\frac{10}{8}$	$\frac{12}{10}$	$\frac{15}{12}$	(larger
Price	20/-	30/-	36/-	45/-	60/-	90/-	to order.)

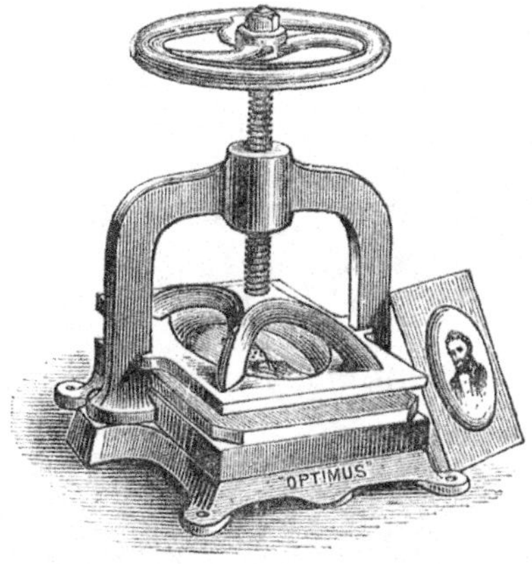

"OPTIMUS" CAMEO PRESSES.

Carte de Visite (3 shapes)	10/-
Cabinet (3 shapes each for Cabinet and Carte de Visite)	22/-

TRADE DISCOUNT LIST ON APPLICATION.

PHOTOGRAPHIC SUNDRIES.

Vulcanite Trays.

	3 by 3½	4¾ by 3¾	5¾ by 4½	7 by 5¼	8 by 6	9 by 7	11¼ by 9¼ by 1	13½ by 11½ by 1½
Price per doz.	5 3	5 9	6/9	8 9	15 -	19/-	27/-	46/-

Porcelain Trays.

	5 by 4	6 by 5	7 by 5	8 by 6	9 by 7	10 by 8	11 by 9	12 by 10	13 by 11	14 by 11
Price per doz.	7/3	8/6	9 3	10 6	12 6	15/-	20/-	25 -	35/-	42/-

Papier Mache.

	4¾ by 3	7 by 5	8¾ by 6¾	10¾ by 8¼	12¼ by 10¼	15⅛ by 12¼
Price per doz.	6 9	11/-	14/-	21 -	28 6	45/-

Folding Plate Racks.

For draining plates after washing: (to hold 12)	10 3	13 6	18/-
,, ,, (,, 24)	14 -	18 -	24/-

Graduated Glass Measures.

	1dr.	2dr.	1oz.	2oz.	4oz	6oz.	8oz.	10oz.	16oz.	20oz.	32oz.	40oz
Per doz.	6/-	7/6	6 9	8 -	10 3	12/9	14/3	15 9	22 3	27/-	40/6	45/-

Glass Funnels, Ribbed or Plain.

Diam of Top Ins.	1½	2	2	3	4	4½	5	5½	6	7	8
Price per doz.	3/-	3/3	3 6	4 0	4/6	5 -	7 -	9/-	12/-	16/-	21/-

Cutting Shapes.

	C.D.V.	4 by 3	Cabinet.	6½ by 4¾	8½ by 6½	10 by 8	12 by 10
Price per doz.	5/6	6 -	9 6	10 6	22 -	36 -	48 -

PHOTOGRAPHIC SUNDRIES.

Printing Frames Oak.

	4¼ by 3¼	5 by 4	6½ by 4¾	8½ by 6½	10 by 8	12 by 10
Superior finish, round corners, brass springs, per doz.	4 9	6 2	8 9	12 6	18 -	27 -
Mahogany, superior, per doz.	9 -	15 -	21 -	32 -	50 -	66 -

Light-tight Plate Boxes.

Mahogany, for 12 plates, per doz.	30/-	43 -	54 0	68 -	78/-	84 -
" " 24 " "	60 -	72 -	80 -	102 6	144/-	180/-

Negative Boxes.

White Wood (to hold 12), per doz.	10 6	12 -	18 -	22 6	37 6	45/-
" " 24), "	12 9	13 6	22 6	28/-	40/-	57/-
" " 50), "	16 6	21/-	28 6	39 -	57/-	66/-

Vignette Glasses.

6 6	12/6	18 -	25/6	30/-	48/-

Carriers or Inner Frames.

For Double Dark Slides, to carry smaller plates, per doz.	15/9	18/-	21 -	27 -	36/-	48/

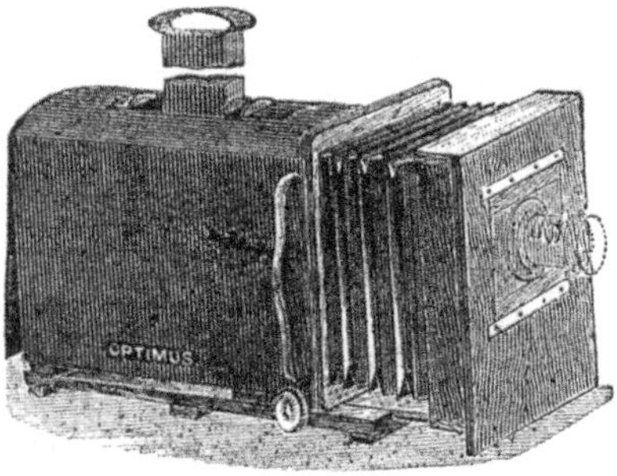

"OPTIMUS" ENLARGING APPARATUS.

This Apparatus comprises superior Ma'iogany Body Lantern and long Bellows Camera adjusted by Patent Quick Action Rack and Pinion. The Lantern is fitted with powerful Refulgent Lamp, with 3 wicks, giving brilliantillumination. Compound Condensers.

CONDENSER.

5-inch	no front lens		...	100/-
5 ,,	with	,,	...	127 -
6 ,,	no	,,	...	133/-
6 ,,	with	,,	...	151/-
7 ,,	no	,,	...	155/-
7 ,,	with	,,	...	173/6
8 ,,	with	,,	...	256/-
9 ,,	with	,,	...	290 -
10 ,,	with	,,	...	360/-
12 ,,	with	,,	...	580/-

If with Russian Iron instead of Mahogany Body.

5inch	no lens		...	75/-
5 ,,	and	,,	...	102/6
6 ,,	no	,,	...	87/6
6 ,,	and	,,	...	115 3
7 ,,	no	,,	...	110 -
7 ,,	and	,,	...	137/9
8 ,,	no	,,	...	160/-
9				210 -
10 ,,	,,	,,	...	285/-
12 ,,	,,	,,	..	450/

Adapted for use with Lime-light or Oil Lamp.

When large sized Condensers are employed, it will be found advantageous to use the Oxy-hydrogen or Oxy-calcium Lime Light Burners; but good results are obtained with the Refulgent Mineral Oil Lamps supplied with the apparatus.

'OPTIMUS'
COMPOUND CONDENSERS (mounted) FOR ENLARGEMENTS.

Inches Diameter	3½	4	4½	5	6	7	8	9	10	12
	8 -	8 6	16 6	24/-	39/-	50 9	69/6	93/6	132/-	222/-

TRADE DISCOUNT LIST ON APPLICATION.

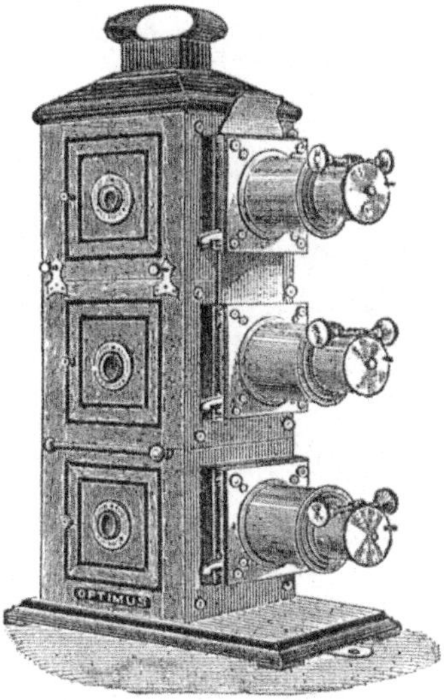

'OPTIMUS'

TRIPLE OXY-HYDROGEN LANTERN

The Top of the Lantern may be used separately with Oil Lamp.

Seasoned Mahogany Body, 6 Panelled Doors with Sight Holes. Moulded foot, picked out with black, Highly Finished Brass Stages and Sliding Tubes, Compound Condensers 4 inches diameter. Three-draw Telescopic Front Tubes, and SIX Photographic Front Lenses of 6-in. and 4-in. focus ...	22	18	0
Three Safety Gas Jets	1	13	0
'OPTIMUS' Triple Dissolving Tap (Patent)	1	13	0
	£26	4	0

The Draw Tubes are specially rigid, so maintaining the Optical Axis accurately and ensuring the Front Lens, Condenser, and Slide occupying Parallel Planes.

Curtain Slide, extra 7/-; Double pinion, extra 3/-; Flashers, extra 3 6 each.
Lantern Photographs, ... Plain, 12/-; Coloured, 18/6 per doz.

TRADE DISCOUNT LIST ON APPLICATION.

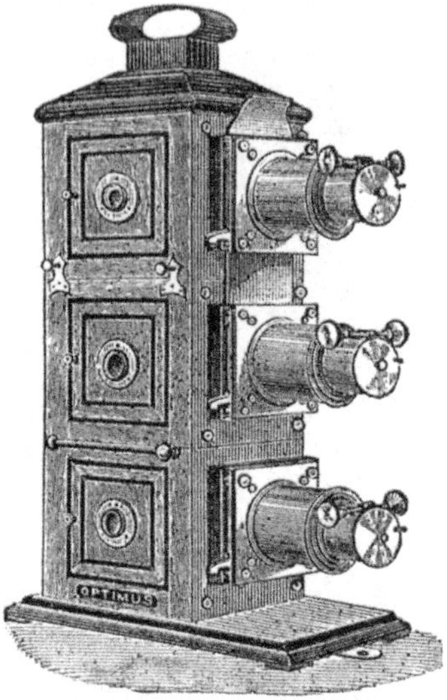

'OPTIMUS'

OXY-HYDROGEN TRIPLE LANTERN.

The Top Lantern may be used separately with Oil Lamp

Seasoned Mahogany body, 6 Panelled Doors with Sight holes. Moulded foot, picked out with black, Finished Brass Stages and Sliding Tubes. Achromatic Photographic Combination Front Lenses, large diameter Back Lens, Compound Condensers of 4 inches diameter	**14**	**10**	**0**
3 Safety Gas Jets	**1**	**13**	**0**
'OPTIMUS' Triple Dissolving Tap (Patent)	**1**	**13**	**0**
Complete	**£17**	**16**	**0**

The Draw Tubes are made specially rigid, so maintaining the optical axis accurately and ensuring the front lens, condenser and slide occupying parallel planes.

Curtain Slide, extra 7/-; Double pinion, extra 3/-; Flashers, extra 3/6 each.
Lantern Photographs, ... Plain, 12/-; Coloured, 18/6 per doz.

TRADE DISCOUNT LIST ON APPLICATION.

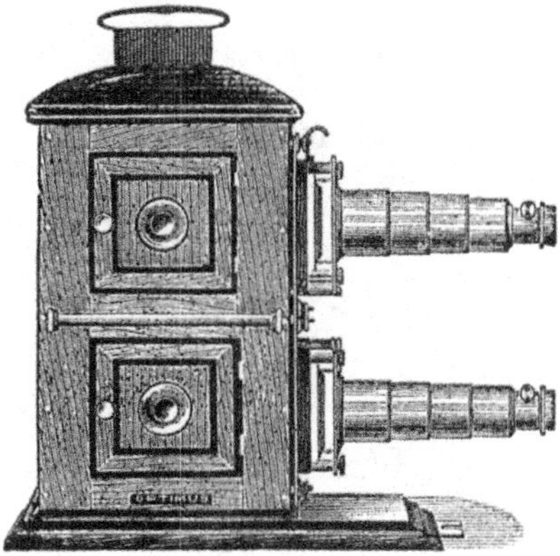

'OPTIMUS' Bi-unial Oxy-Hydrogen LANTERN.

Seasoned Mahogany Body, 4 Panelled Doors with Sight Holes, Moulded Foot, picked out with Black, Compound Bi-convex Condensers of 4 inches diameter. Highly Finished Brass Stages, and with Brass 3-draw Telescopic Front Tubes, and FOUR Photographic Front Lenses of 6 inches, 4 inches focus	£14 14s.
2 Safety Jets	£1 2s.
6 Way Dissolving Tap	17s.
	£16 13s.

The Draw Tubes are specially rigid, so maintaining the Optical Axis accurately, and ensuring the Front Lens, Condenser, and Slide occupying Parallel Planes.

Curtain Slide, extra 7/-; Double pinion, extra 3 -; Flashers, extra 3 6 each.
Lantern Photographs, ... Plain, 12 -; Coloured, 18 6 per doz.

TRADE DISCOUNT LIST ON APPLICATION.

PERKEN, SON, & RAYMENT, 99, Hatton Garden, LONDON.

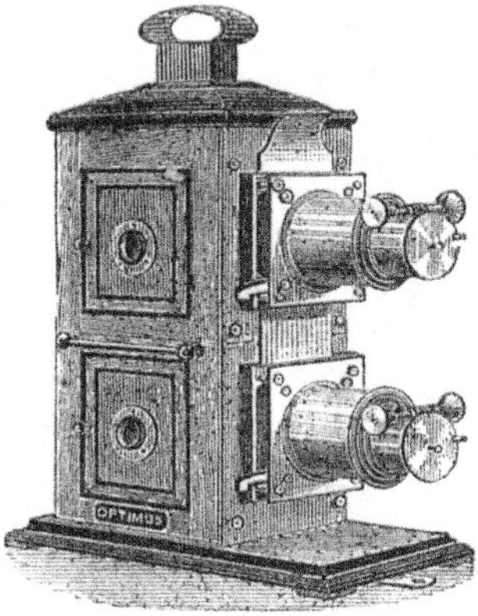

'OPTIMUS' BI-UNIAL LANTERN

For Oxy-Hydrogen Lime Light.

The Top Lantern may be used with Oil Lamp.

Seasoned Mahogany Body, 4 Panelled Doors with Sight Holes, moulded foot picked out with black, Japanned stages and tubes, Achromatic Photographic Front Lenses, compound condensers of 4 inches diameter	**£5 12 0**
Ditto ditto with highly finished BRASS stages and sliding tubes...	**8 8 0**

The Draw Tubes are specially rigid, so maintaining the Optical Axis accurately and ensuring the Front Lens, Condenser, and Slide occupying Parallel Planes.

Curtain Slide, extra 7/-; Double pinion, extra 3/-; Flashers, extra 3/6 each.
Lantern Photographs, ... Plain, 12/-; Coloured, 18/6 per doz.

TRADE DISCOUNT LIST ON APPLICATION.

'OPTIMUS' SIDE-BY-SIDE OR BI-UNIAL MAGIC LANTERN.

May be used as a pair with Oil or Lime-light.

The above Diagram represents a full-sized combined Lantern. It is made of japanned metal. It may be worked one above the other, as the dotted lines show, or side by side as the positive diagram shows: or again, the two Instruments may be separated and worked in two distinct places, as each Lantern is complete in itself. A further advantage is possessed by these Lanterns, for the body which is constructed to accommodate limelight will also readily accommodate oil lamps, the groove into which the trays are inserted being made to the same guage as our Lamps.

Without Lamps or Jets		**£4 16 0**
Gas Jets	each...	**11 0**
8-Wick Lamps, each		**12 0**

Achromatic Photographic Combination Front Lenses (with large diameter Back Lens). Compound Condensers of 4 inches diameter.

Curtain Slide, extra 7/-; Double pinion, extra 3/-; Flashers, extra 3/6 each.
Lantern Photographs, ... Plain, 12/-; Coloured, 18/6 per doz.

TRADE DISCOUNT LIST ON APPLICATION.

'OPTIMUS' MAGIC LANTERN.

Japanned Metal Body **30/-**

Adapted for use with Limelight. For Dissolving 2 Lanterns are necessary.

STUDENT'S MAGIC LANTERN.

Students Lantern Japanned Metal Body (to take demonstrating tank) with finished Brass Sliding Tubes **40/-**

Do. Russian Iron Body, with finished brass front stage plate and sliding tube **55/-**

Adapted for use with Limelight. For Dissolving, 2 Lanterns are necessary.

Each Magic Lantern is efficient for exhibitions. The Lens gives crisp definition, being a superior Achromatic Photographic Combination (large diameter back lens). with rack and pinion. It is fitted to a telescopic lengthening tube, so gaining increased focal accommodation. The Condenser is composed of two plano-convex lenses of 4 inches diameter The refulgent lamp has 3 wicks, or 4 wicks 2s. extra, yielding a brilliantly illuminated picture.—Each is complete in box.

RUSSIAN IRON MAGIC LANTERN.

Highly Finished Brass Sliding Tubes ... . **45s.**

Adapted for use with Limelight. For Dissolving, 2 Lanterns are necessary.

Each Magic Lantern is efficient for exhibitions. The Lens gives crisp definition, being a superior Achromatic Photographic Combination (large diameter back lens), with rack and pinion. It is fitted to a telescopic lengthening tube, so gaining increased focal accommodation. The Condenser is composed of two plano-convex lenses of 4 inches diameter. The refulgent lamp has 3 wicks, or 4 wicks 2s. extra, yielding a brilliantly illuminated picture.—Each is complete in box.

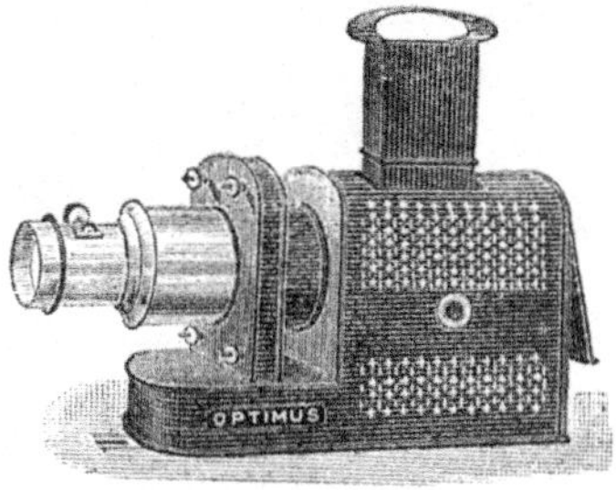

"OPTIMUS" MAGIC LANTERNS.

Perforated Russian Iron Body, Brass Sliding Tubes ... 50s.

Adapted for use with Limelight. For Dissolving 2 Lanterns are necessary.

Each Magic Lantern is efficient for exhibitions. The Lens gives crisp definition, being a superior Achromatic Photographic Combination (large diameter back lens), with rack and pinion. It is fitted to a telescopic lengthening tube, so gaining increased focal accommodation. The Condenser is composed of two plano-convex lenses of 4 inches diameter. The refulgent lamp has 3 wicks, or 4 wicks 2s. extra, yielding a brilliantly illuminated picture.—Each is complete in box.

TRADE DISCOUNT LIST ON APPLICATION.

"OPTIMUS" MAHOGANY MAGIC LANTERN.

Mahogany outside Body, Japanned Metal Stages and Sliding Tubes ... **42s.**

"OPTIMUS" SUPERIOR MAHOGANY MAGIC LANTERN.

Seasoned Mahogany Body, 2 Panelled Doors, Highly Finished Brass Stages and Sliding Tubes ... ... **82s.**

Adapted for use with Limelight. For Dissolving, 2 Lanterns are necessary.

Each Magic Lantern is efficient for exhibitions. The Lens gives crisp definition, being a superior Achromatic Photographic Combination (large diameter back lens), with rack and pinion. It is fitted to a telescop'c lengthening tube, so gaining increased focal accommodation. The Condenser is composed of two plano-convex lenses of 4 inches diameter. The refulgent lamp has 3 wicks, or 4 wicks 2s. extra, yielding a brilliantly illuminated picture.—Each is complete in box.

TRADE DISCOUNT LIST ON APPLICATION.

'OPTIMUS'
PAIRS OF LANTERNS FOR DISSOLVING.

For Dissolving Views, two Lanterns are required, and are arranged in one portable box; either of the before mentioned styles of Lanterns may be selected. The extra cost above that of the two lanterns being 8/9, the price of the fan dissolver; if used for oxy-hydrogen light two gas jets are needed.

Japanned Metal Body	30 -	each.
,, ,, Student's Form	40/-	,,
Mahogany Body, Lined with Metal	42 -	,,
Russian Iron Body	45/-	,,
,, ,, Perforated	50 -	,,
,, ,, Student	55 -	,,
Seasoned Mahogany Body, 2 Panelled Doors, all Brass Fittings ...	**82/-**	,,

Each Magic Lantern is efficient for exhibitions. The Lens gives crisp definition, being a superior Achromatic Photographic Combination (large diameter back lens), with rack and pinion. It is fitted to a telescopic lengthening tube, so gaining increased focal accommodation. The Condenser is composed of two plano-convex lenses 4 inches diameter. The refulgent lamp has 3 wicks (or 4 wicks, 2s. extra), yielding a brilliantly illuminated picture.—Each is complete in box.

Safety Blow through Gas Jets	11/-	each.
Chamber (mixed) ,, ,, ...	16/6	,,

Lantern Photographs, Plain, 12 -; Coloured, 18 6 per doz.

TRADE DISCOUNT LIST ON APPLICATION.

'OPTIMUS' MAGIC LANTERNS FOR YOUTHS.

To burn Paraffin or Mineral Oil.

The body of these lanterns is so constructed that the oil reservoir is not likely to become heated, since it falls through the bottom of the illuminated chamber and is in outside air.

Small Magic Lanterns, with condensers, front lens (adjustable), black japanned body, chimney, lamp and reflector—

No. 1	...	diam. front lens	$1\frac{1}{8}$	...	condenser	$1\frac{3}{4}$	... 2'6 each
2	...	"	$1\frac{1}{4}$	...	"	$1\frac{3}{4}$	... 3 10 "
3	...	"	$1\frac{5}{16}$	...	"	$2\frac{1}{16}$	... 6/6 "
4	...	"	$1\frac{1}{2}$	...	"	$2\frac{5}{8}$	... 9/- "
5	...	"	$1\frac{1}{2}$	...	"	3	... 11- "
6 (in box)		"	$1\frac{5}{8}$	...	"	$3\frac{1}{2}$	... 20- "

Boxes of 12 Slides for Lantern, each Slide containing several figures arranged as Tales if desired.

No. 1	Paper edge	3 - each.
2	"	3 8
3	Wood edge	6 8
4	"	9 4
5	"	12 -
6	"	16 -
7	"	18 8

NOTE.—We maintain the sizes, both of Lanterns and Slides, as of old: many makers call our No. 2 No. 3, and so on.

TRADE DISCOUNT LIST ON APPLICATION.

RACKWORK & MECHANICAL SLIDES.

For Lanterns with 4 inch Compound Condensers only.

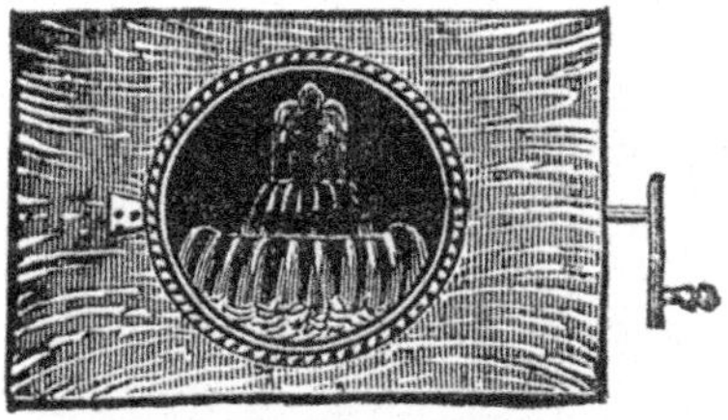

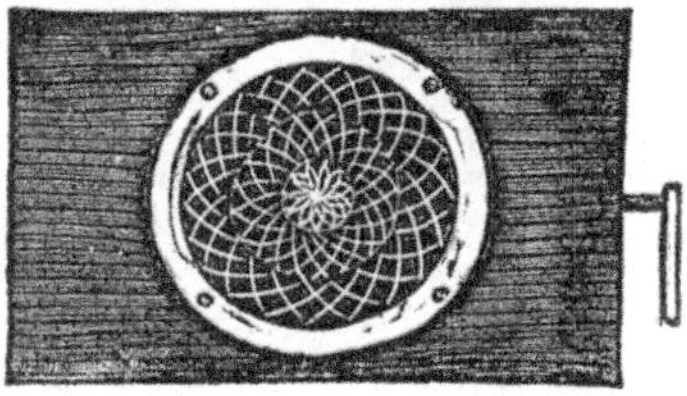

	s.	d.	
CHROMATROPES, 3 inch diameter, best quality (various designs)	9	0	each.
CHROMATROPES, 3 inch diameter, with Photographic Portraits, Mottoes, Views, &c. in centre	10	0	,,
CHROMATROPE CARRIER FRAME, with 6 Pairs of Painted Discs in rack box	21	0	,,
RACKWORK SOLDIERS' HEADS, changing to Donkeys' Heads	8	6	,,
SNOWSTORM FFFECT	7	0	,,
CURTAIN to roll up, showing Painted Curtain or Drop Scene ...	10	0	,,
DANCING SKELETON SLIDE, with lever motion taking oft Head	5	0	,,
LIGHTNING EFFECT	1	9	,,
RAINBOW EFFECT	1	9	,,
LEVER MOON-RISING EFFECT	4	6	,,
MAN SWALLOWING RATS, rack mills, &c.	10	0	,,
RACK WORK, best painted wave slides	15	0	,,

TRADE DISCOUNT LIST ON APPLICATION.

SLIDING CARRIER BLOCK.

	s.	d.
For two pictures, per doz.	27	0
SETS OF EFFECT SLIDES for dissolving views from... per set	4	6
Sets of all the most popular Nursery Tales, in great variety best painted ... ... per slide...	4	0
Second quality ... ... per slide...	2	8
LONG PANORAMA SLIDES ... ... from...each...	2	3
Motto Slides ... ... ,, ,, ...	3	0
SETS OF 10 SLIDES, best rack Astronomy, 2½-inch circles, in mahogany box	100	0
Ditto ditto in 3-inch ,, ,,	110	0
PHOTOGRAPHIC VIEWS taken in all countries per dozen	12	0
Colored ditto ... ... from per dozen	18	6
SQUARE WOOD BLOCKS, per gross	42	0
Soft limes, per tin of one dozen	1	0
Hard limes	1	6
Front lenses for lantern, rack motion, double achromatic photographic combination...	15	0
Ditto ditto long focus	36	0
,, ,, ,, ,, ...	46	0
GAS JETS.		
Blow Through	11	0
Chamber	16	6
Interchangeable	18	6

GAS BAGS, GAS CYLINDERS,

At lowest market prices.

SOLE AGENTS FOR THE

PATENT FANTOCCINI ANIMATED TRANSPARENT FIGURES

Which without doubt give a most life-like effect when projected on the screen.

MAGIC LANTERN SLIDES IN GREAT VARIETY.

TRADE DISCOUNT LIST ON APPLICATION.

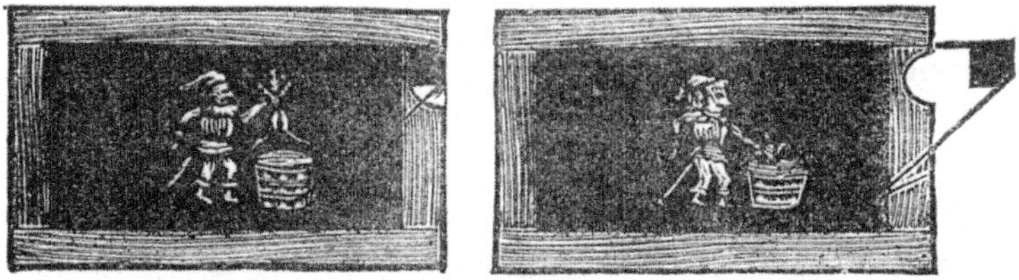

COMIC SLIPPING SLIDES.

600 Different Subjects ... ... per doz. **12/-**

These are well painted, giving brilliancy and transparency of colour.

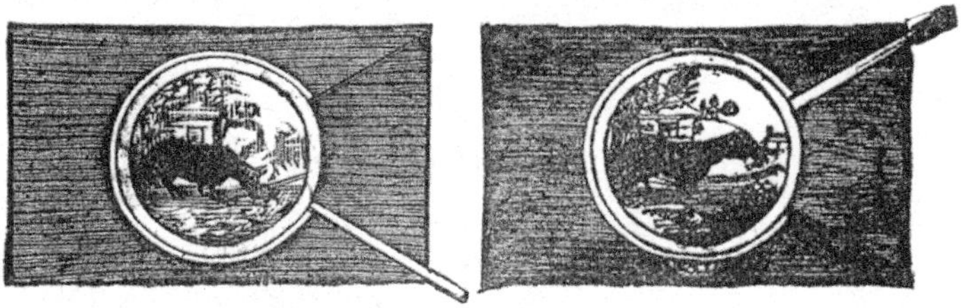

COMIC LEVER SLIDES.

300 Different Subjects ... ... per doz. **37/6**

These mechanical slides are suited for Lanterns with 4-inch condensers.

SLIDING CARRIER BLOCK, for two pictures, per doz. ... **27/-**

TRADE DISCOUNT LIST ON APPLICATION.

PERKEN, SON, & RAYMENT, 99, Hatton Garden, LONDON.

MICROSCOPE FOR LANTERN.

Having brass body with high and low powers, suitable for optical lantern possessing 4in. compound condensers 18/6

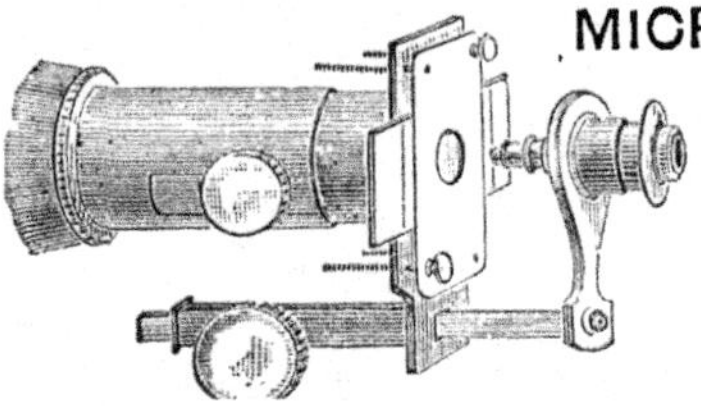

MICROSCOPE FOR LANTERN.

(SUPERIOR.)

If with rack motion and revolving stage 84/-

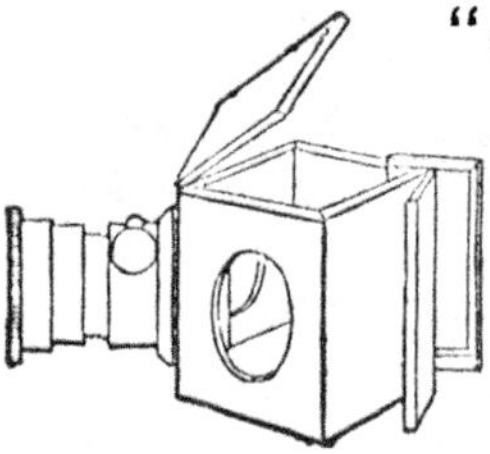

"OPTIMUS" APHENGESCOPE.

An instrument for exhibiting opaque objects, cartes-de-visite, etc.; suitable for optical lantern with 4in. compound condensers 16/-

Superior Aphengescope, arranged for pairs of lanterns with achromatic front lenses and rack adjustments 30/-

"OPTIMUS" POLARISCOPE.

With analyser in case 135/-

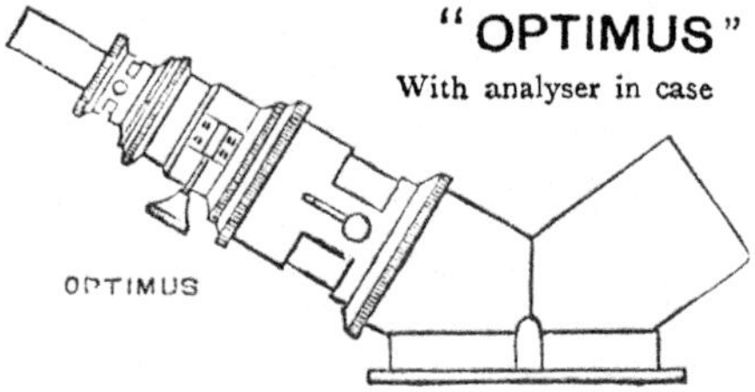

"OPTIMUS"

KALEIDOSCOPE.

In case with lock and key 40/-

TRADE DISCOUNT LIST ON APPLICATION.

"OPTIMUS" REFULGENT LAMPS.

Stout Russian Iron. Burning Mineral Oil.

This lamp has been universally pronounced the *safest* yet the *most brilliant* lamp used with the Projecting Magic Lantern. A flame-sight-hole at the back enables the manipulator to adjust the wicks with the nicest accuracy whilst working.

Two wicks (wicks 2 inches wide)	12 6
Three wicks	13 6
Four wicks	15 6

"OPTIMUS" LAMP WICK TRIMMER (Patent).

For 1½ inch wicks	2/6
For 2 inch wicks	4 9

This little contrivance saves much trouble. With one cut it cleans off the charred portion of wick, and leaves a perfectly even ridge of cotton to hold the flame, which, in consequence, burns with a clearness and freedom from smoke that is very desirable. It is particularly recommended for photographic developing lamps, and for lamps used with the optical lantern.—*British Journal of Photography.*

ANIMALCULÆ OR CHEMICAL TANK.

These being made of Glass and Indiarubber, are not affected by chemicals **3 6**

DISCOUNTS ON APPLICATION.

MAGIC LANTERN GAS JETS FOR LIMELIGHT.

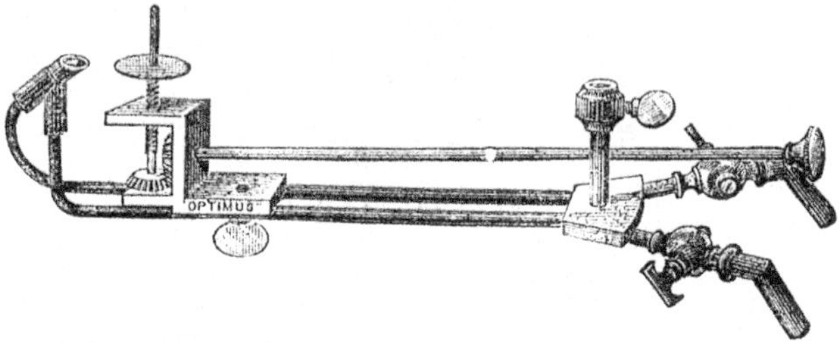

Be·t quality blow-through gas jet, with cog-wheel arrangement for turning, also raising and lowering the lime; platina nipple ... ... each **11/-**

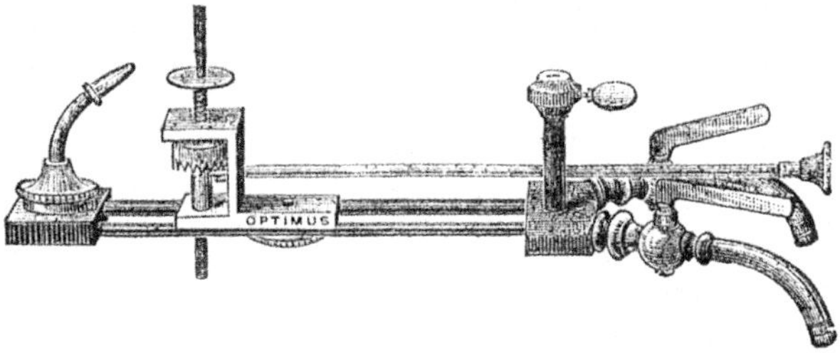

Best mixed chamber gas jet, with cog-wheel arrangement for turning, also raising and lowering the lime; platina nipple ... ... each **16/6**

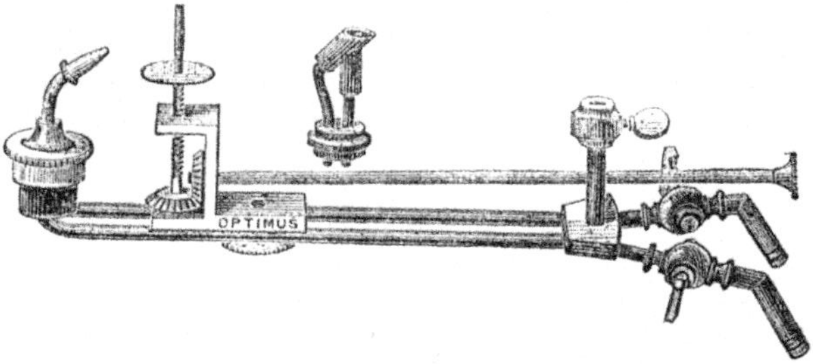

New interchangeable jet or both gases under pressure, mixed or for blow-through form, by simply removing and using the burner as required; both have platina nipples ... ... each **18/9**

TRADE DISCOUNT LIST ON APPLICATION.

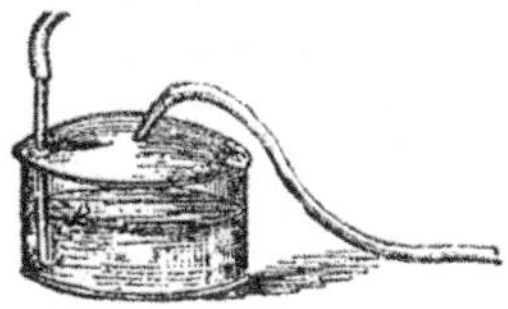

ZINC PURIFIER, 4/-

BRAZED IRON RETORT, 10/-

GAS CYLINDER

With Regulator and Pressure Gauge attached.

GAS BAGS—Stout Twill.

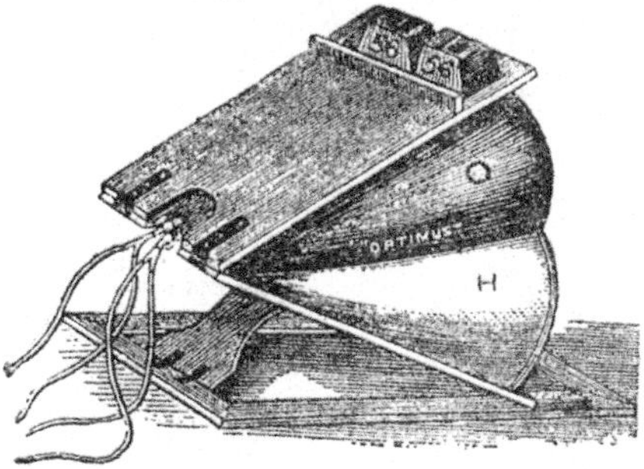

Subject to Fluctuations.

SIZES. in.	in.	in.	Capacity cubic ft.	Pric
30 by	24 by	20	.. 4	.. 32
36	24	20	.. 5	.. 37 6
36	24	24	.. 6	.. 39 6
36	28	24	.. 7	.. 42 -
36	32	24	.. 8	.. 45/-
36	36	24	.. 9	.. 46 6
40	32	24	.. 9	.. 48 6
40	36	24	.. 10	.. 51/-
40	36	26	.. 11	.. 53/-
40	36	28	.. 12	.. 55/-

TRADE DISCOUNT LIST ON APPLICATION.

PATENT SAFETY POROUS

ETHER SATURATOR,

FOR PRODUCING THE

ETHO-OXYGEN LIME LIGHT,

FOR OPTICAL LANTERNS.

This Apparatus is the **most perfect** means for producing the Etho-Oxygen Lime Light, and gives the most powerful illumination known for Optical Lanterns. In its use there is no heat, no bubbling, and no obstruction to the free passage of the Oxygen. It can be disturbed or upset without affecting the light or spilling any Ether into the tubing. After one adjustment of the light, it will automatically regulate the supply of Ether vapour to correspond to any variation in the supply of Oxygen. **This is a very important advantage possessed by no other Apparatus.**

ILLUSTRATION.

It consists of two brass tubes or bodies, screwed into a bent brass connecting tube, lying side by side on an ebonised wooden stand, which is fitted with a handle for carrying about. The brass tubes or bodies can easily be removed by unscrewing the set screws which fasten them to the board. Each tube is fitted at one end with screw-down cap, having a nozzle in its centre for the purpose of attaching elastic tubing, and both are fitted with a roll of flannel or coarse cloth, having a spiral wire in the centre to keep them open. Each nozzle has a small screw cap to prevent escape of Ether when not in use.

PRICE 36-

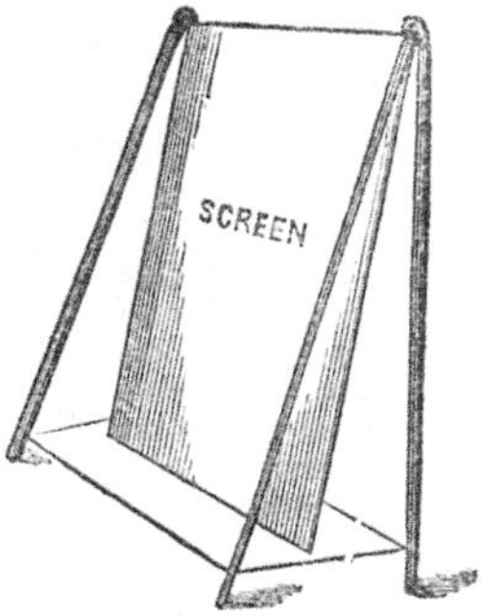

TRANSPARENT SCREENS. Best Union Cloth.

Feet square		6	7	10	12	14	16	20
	27-	28-	30-	36/-	45/-	60-	75/-	126-

"OPTIMUS" (LECTURER'S) PORTABLE READING LAMP.

Price of Lamp, with Signal, Bell, and Match Box complete ... ... 8/- each

---

TRADE DISCOUNT LIST ON APPLICATION.

PERKEN, SON, & RAYMENT, 99, Hatton Garden, LONDON.

SETS OF LANTERN

PHOTOGRAPHS.

Price Uncoloured	.. ..	12s. per dozen.
„ Coloured	.. ..	18s. 6d. „

Number of Slides comprised in each Set is stated in figures attached to each Title.

A

A Day in London .. ... set of	60
A Day's Holiday at Windsor ...	30
A Photographer's Perplexities ...	12
A Precious Couple on the road to Gretna Green	12
A Trap to Catch a Sunbeam ...	15
A Trip to Brighton...	40
A true Story of the old Coaching Days	10
A Visit to the British Museum ...	50
A Walk in the "Zoo"	48
A Year within the Arctic Circle...	50
Abbeys and Castles of England ..	50
Adventures of Briggs with a Bull	4
Adventures of Brown, Smith, Jones and Robinson	4
Æsop's Fables	50
Afghan War...	36
Aladdin, or the Wonderful Lamp	1
Algiers	20
Alice in Wonderland	42
„ „ Songs ...	7
America, North, and Niagara ...	50
American Franklin Search Expedition	20
Androcles and the Lion	6
Animal Sagacity	24
Animals at the Zoological Gardens	215
An Old Story	26
Art Collections, South Kensington Museum	50
Astronomy ... ... } Astronomical Instruments }	300

B

Baby's Sugar Bag	12
Barber and his Clever Dog ...	12
Baron Munchausen	8
„ „	17
Barred Window	8
Bashful Man...	6
Beautiful Snow	5
Belgium	50
Berlin...	46
Bible Manners and Custom ...	30
Blunderb re	12
Book of Martyrs	52
Bookworm...	12

B (*continued*).

Boons and Blessings	16
Bottle, The	8
Boys and Ravens	8
Boys of Corinth	12
Brazilian Ape	14
Brighton Aquarium	14
Briggs and the Bull	78
British Museum, Antiquities of &c.	28
Brown and the Bear	12
Brown and Mouse	12
Brussels	50
Bunyan, Life of	111
Bunyan's Pilgrim's Progress "Art Journal"	22
Do. Cassell	36
Bunyan's Pilgrim's Progress Religious Tract Society, set of	17
Do. Routledge	111
Do. Service of Song	33
Do. Christiana	32
Burglars, The Two..	12
Burmah	36
Buy your Own Goose	6
Buy your Own Cherries	7 & 10

C

Canada from Quebec to the Rocky Mountains	60
Cats and Serpents	8
Cathedrals, Our English	50
Cat, Nine Lives of a	12
Central Africa	50
Children in the Wood	8
Channel Islands	60
Children's Entertainment, 1 ...	48
„ „ 2 ..	48
Children's Messiah, Service of Song	32
Chimes, The	24
China and the Chinese	60
Christmas Set	17
Christie's Old Organ ...	24 & 17
Cinderella	8
Cleopatra's Needle...	47
Clever Nephew	12
Cock Robin	8
Come Home, Father	8

Price of above Lantern Photographs: Plain 12/-, Coloured 18/6 per doz.

TRADE DISCOUNT LIST ON APPLICATION.

SETS OF LANTERN PHOTOGRAPHS.

C (*continued*).

Comic Slipping Slides	100
Cornwall	50
Crochet's Musical Recital... ..	12
Cruikshank's Works	28
Crusoe, Robinson	17
"Curfew must not ring to-night"	10
Colony of New Zealand. Part 1	54
" " " 2	30

D

Dame Perkins and her Grey Mare	8
Dan Dabberton's Dream	14
Day in London	60
Dear Father, Come Home ...	3
Death of the Bluebottle	12
Devonshire	50
Dickens' Chimes	24
Dick Whittington	21 & 8
Diogenes and the Boys of Corinth	12
Donkey and Mill	12
Doré's Bible Illustrations... ...	25
Dr. Spiritus and the Moon ...	6
Dreams at Sea	5
Drunkard's Children	8
" Progress	14
Ducks and Frog	12

E

Egypt... ... . set of	60
Egypt, Modern, and its People ...	50
Egyptian War, The	50
Elephant's Revenge	12
Engadine Switzerland, The ...	50
English Cathedrals...	50
English Lakes	50
English River Scenery	50
Eva	26

F

Fables, Æsop's	50
" La Fontaine's	135
Five Senses	5
Flea, Adventures with a	12
Fly, Adventures with a	12
Florence, The City of	30
Foolish Toper	12
Four Seasons	4
Foxe's Book of Martyrs	52
Friendless Bob	18

G

Gabriel Grub	17
General Description and Statistics of London	48
General Gordon	24
Gilpin, Johnny	12
Gin Fiend	4
Gin Shop	12
Golden Goose	12
Gossips	12
Greatest Plague of Life	12
Groups and Ideal Photos. from Life	83

H

Haddon Hall	18
Hardanger Fjord Norway, The...	41
Heat	200
Heathen Chinee	9
Highlands of Scotland, The ..	52
Holland	50
History of a Pound of Tea ...	10
" a Cotton Bale	10
" a Quartern Loaf ...	10
" a Pound of Sugar ...	10
" a Golden Sovereign ..	10
" a Scuttle of Coals ...	10
Hogarth's Works	40
" Harlot's Progress, The	6
" Industry and Idleness	12
" Marriage à la Mode ...	6
" Rake's Progress, The	8
" Reward of Cruelty ...	4
" Rumours of an Election	4
Holy Land	60
Honey Stealers	8
Hostile Neighbours	12
Housebreakers, The	12
Human Physiology	52
Hunting Expedition, H.R.H. the Prince of Wales ..	30 & 30
Hymns, Christie's Old Organ, set of	17
Hymns, Words only	300

I

Impulsive Gardener	8
India, Mysore	51
Introductory Slides to various subjects	46
Introductory Slides and Mottoes	43
Inverted World	6
Ireland, 1	60
" 2	50
Italy	50
Italian Lakes	50

Price of above Lantern Photographs: Plain 12/-, Coloured 18/6 per doz.

SETS OF LANTERN PHOTOGRAPHS.

J, K

Jack and the Beanstalk	8
Jack the Conqueror	12
Jack the Giant Killer	8
Jackdaw of Rheims	13
Jane Conquest	16
Jessica's First Prayer ...	10 & 18
Jocko the Brazilian Ape	14
John Hampden's Home	6
Johnny Gilpin	12
John Ploughman's Pictures ..	38
John Ploughman's Picture Hymns	17
John Tregenoweth: his Mark ...	18
Jones' Baby	4
Joseph, Service of Song	33
Juvenile Smokers	6
Kate Maloney	6

L

La Fontaine's Fables	135
Lady Jane Grey	4
Lakes, English	50
,, Italian	50
Lazy Traveller	12
Level Crossing	9
Life Boat	7
Life of Martin Luther	12
Light	300
Little Artist and Large Portfolio	8
Little Jim the Collier Boy ...	6
Little Red Riding Hood	8
Little Tiz	14
Little Town of Weinsburgh ...	6
Liverpool	42
London Street Traffic	79
London and Neighbourhood ...	932
London to Rome	50
London to the Falls of Niagara...	46
Lovechase and his Dog Tray ..	12

M

Mad Umbrella	12
Marley's Ghost, a Christmas Carol	25
Mary, Queen of Scots	24
Mary, the Maid of the Inn ...	10
{ Magnetism and Electricity ...	33
{ Magnetic Curves	60
Man and Calf	12
Maps .. .. set of	21
Martin Luther	12
Mechanics, Hydrostatics, &c. ...	110
Mediterranean	50
Meg and her Brother Ben ...	13
Messiah, Service of Song	32
Microscopic Gems	50
Microscopical Objects	150
Midnight Adventures with a Flea	12

M (*continued*).

Miller and the Sweep	12
Mines and Miners	130
Mines and Mining	41
Miss Pop's Pet	12
Mistletoe Bough	7
Modern Egypt	50
Morrow of the Carouse	8
Mother's Last Words ... ...6 &	12
Mottoes and Texts...	140
Mouse, Midnight Adventures with	12

N

Nelly's Dark Days	14
New Arctic Expedition	9
New Hat	9
Newton, Sir Isaac, and the Apple	8
New York	50
New York to the White Mountains	60
Niagara	59, 41, 46
Nine Lives of a Cat	12
North American Scenery	148
Norway	40

O

Oiled Feather	12
Old Curiosity Shop, The... ...	24
Old Man and his Ass	7
Old Mother Hubbard	8
Old Story	26
Old Testament Scriptures, Reading	40
Origin of Jones' Baby	4
O'Toole and the Umbrella ...	9
Outcast London; or, How the Poor Live	40
Our West African Settlements ...	48
Oxford Colleges	78
Oxford to London Bridge... ...	50

P

Parasites and their Hosts... ...	40
Paris	50
Passions, The	12
Passion Play	50
Peasant and his Ass	7
Peep into Nature through the Microscope, A	54
Phonograph	134
Photographic Sketches of English Life and Scenery	40
Photograph of a Lightning Flash	1
Plant Life	39
Portraits	39

Price of above Lantern Photographs: Plain 12/-, Coloured 18/6 per doz.

SETS OF LANTERN PHOTOGRAPHS.

P (*continued*).

Power of Music ... ... set of	6
Prince of Wales's Animals at the Zoo	17
" " Hunting Expedition at Nepaul	30
Prodigal Son, Service of Song ...	38
Professor Crotchet's Musical Recital	12
Progress of Intemperance ...	6
Pussy's Road to Ruin	12

Q

Quarrelsome Dog	12
Quartette Party	9
Queen's Jubilee, 1887	50

R

Retaliation	12
Return from the Tavern, The ...	4
Reward of Covetousness ...	8
Reward of Cruelty	4
Reynard the Fox	12
Rhine...	60
Rip Van Winkle	12
Rival Lovers	12
Riviera, The...	50
Robinson Crusoe	17
Robbers	12
Romance of History	44
Rome	50
Rome, Ancient and Modern ...	50
Round the World with a Camera	60
Round the World in a Yacht ...	45

S

Sambo's Five Senses	5
Sankey's Hymns (Words only) ...	76
Sayings of Jesus, Service of Song	29
Scripture, Old Testament ...	154
" " Joseph	124
Scripture, New Testament ...	77
Scripture Texts	76
Scrub, the Workhouse Boy ..	11
Seasons	4
Serpents and the Cat	8
Service of Song, Eva	26
" " Joseph	33
" " Messiah... ...	32
" " Prodigal Son ...	38
" " Pilgrim's Progress	33
" " Sayings of Jesus	29

S (*continued*).

Seven Ages of Man	7
Signal Box	6
Simon and his Pig	12
Sir Isaac Newton and the Apple	8
Sir John the Giant Slayer ...	24
Sketches, English Life and Scenery	40
Sledge Party	12
Sleepy Hollow	6
Snow-white	7
Scotland	102
Solar System	50
Soldier's Dream, The ... set of	8
Soudan War...	57
Sound...	177
South Kensington Museum, Art Collections	59
Spain	50
Spectrum Analysis..	100
Stanley in Africa	29 & 29
Statuary	78 & 113
Statuary in South Kensington Museum	59
Stolen Sausage	9
Sultan of Ragobaga	15
Suspicious Travellers	12
Switzerland, Tour No. 1	50
" " No. 2	50

T

Tabernacle in the Wilderness ...	15
Tale of a Tub, Set of	7
" " Set of	12
Temperance Mottoes	93
The Chimes	24
The Cottar's Saturday Night ...	9
The Jackdaw of Rheims	13
The Knight and the Lady ...	6
The Old Curiosity Shop	24
The Pied Piper of Hamelin ...	12
The Temperance Sketch Book ...	32
The Three Bears	8
The Vagabonds	10
The Well of St. Keyne	4
The Witch's Frolic...	12
The Life Boat	7
The Quaker and the Robber ...	4
The Signal Box	6
The Quartette Party	9
The Baby's Sugar Bag	12
The Rhine	60
The Women of Mumbles Head ...	8
Three Kittens that lost their Mittens	9
Through Turkestan to the Afghan Frontier... ... ... set of	50

Price of above Lantern Photographs: Plain 12/-, Coloured 18/6 per doz.

SETS OF LANTERN PHOTOGRAPHS.

T (*continued*).

Tipsy Geese	6
Toothache	12
Tower of London, The	20
Travels of the Sultan of Ragobaga	15
Trap to Catch a Sunbeam ...	15
Trial of Sir Jasper	25
Two Boys and Raven's Nest ...	8
Two Housebreakers	12
Two Months in India with a Camera	40

U, V

Umbrella, O'Toole Adventures ..	9
Unskilful Ratcatchers	12
Underground Life	90
Venice	30
Vulgar Boy	6
Voyage of the "Challenger" ..	42

W, X, Y, Z

Wales, North	50
Wanderings in Paris	50
Washington City	50
,, to Yellowstone Park	50
Weather Prospects	6
Wedding Bells	10
Westminster Abbey	36
Whisky Demon	12
Whittington and his Cat	24
Witches Frolic	12
Woodman, The Little	12
World Inverted	6
Worship of Bacchus	14
Wreaths and Mottoes	36
Wye River	45
Yacht, Round the World in a ...	45
Zoological Gardens, The Animals	215
,, ,, H.R.H. The Prince of Wales's Animals ...	18

PRICE OF

THE FOREGOING LANTERN PHOTOGRAPHS.

PLAIN	12 -
COLOURED	18/6

SIX PENCE.

Third Edition Revised and Enlarged.
24th Thousand.

BEGINNER'S GUIDE TO PHOTOGRAPHY.

FOR PRESS OPINIONS SEE OVER.

PRESS OPINIONS.

Third Edition. Beginner's Guide to Photography, **6d.**

GRAPHIC.

"The 'Beginner's Guide to Photography' (Perken, Son and Rayment), by a 'Fellow of the Chemical Society,' is a useful little manual for amateur photographers. It contains brief and concise directions for taking, developing, and printing the negative, while there is a valuable article on that bugbear of all amateurs—'Exposure,' by Mr. A. S Platts, containing some exceedingly useful exposure tables "

DAILY NEWS.

"Under the title of the 'Beginners Guide to Photography,' by a 'Fellow of the Chemical Society,' Perken, Son and Rayment have published a useful handbook for all interested in the art of photography. An article on 'Exposure,' and some carefully compiled exposure tables, by Mr. A. S. Platts, must be of value to all amateurs."

St. STEPHEN'S REVIEW.

"'Beginner's Guide to Photography,' published by Perken, Son and Rayment, 99, Hatton Garden, London.—The fashionable art science, Photography, is most explicitly set forth without the confusing technicalities employed in most works on this subject. The difficult matter of 'Choice of Apparatus has a chapter devoted to it, in which the special advantages of each kind of camera and lens is detailed. Altogether this book may be said to be of the greatest value to all who practise photography.

ILLUSTRATED SPORTING & DRAMATIC.

"The 'Beginner's Guide to Photography.'—With this title a six penny book has been published by Messrs. Perken, Son and Rayment, of 99, Hatton Garden, which we find both simple and practical. By following its instructions carefully the amateur will save much disappointment in the sense of blurred pictures, and much expense for spoilt plates."

MORNING POST.

"The 'Beginner's Guide to Photography' is one of the best works on this popular and fascinating art yet published. The author thoroughly understands his subject. Messrs. Perken, Son and Rayment, Hatton Garden, are the publishers."

WHITEHALL REVIEW.

"'Beginner's Guide to Photography.' (Perken, Son and Rayment.)—This is an excellent treatise which all amateurs who have taken up photography as an amusement should peruse."

ARMY & NAVY GAZETTE,

"Messrs. Perken, Son and Rayment send us the second edition of their 'Beginner's Guide to Photography,' a plain and practical handbook as to how to buy and use a camera, with many particulars concerning lenses and other matters, for which the publishers are celebrated as makers."

COURT JOURNAL.

"Messrs. Perken, Son and Rayment, one of the largest and most popular makers of photographic apparatus, publish a most useful little work entitled, 'Beginner's Guide to Photography,' in which the several operations of taking, developing and printing the photograph are described with great clearness, and in a manner suitable to those who are handling a camera for the first time. While those who have not yet provided themselves with the necessary apparatus cannot do better than peruse the valuable chapter on 'The Choice of Apparatus,' and patronise this firm for their purchases."

JEWELLER & METALWORKER.

"'Beginner's Guide,' published by Messrs. Perken, Son and Rayment, of 99, Hatton Garden, at the small sum of six pence. It is a work which can be relied upon, and the language of it is easy of comprehension, a great merit in works of this description."

ILLUSTRATED LONDON NEWS.

'The 'Beginner's Guide to Photography,' published by Messrs. Perken, Son and Rayment, of Hatton Garden, treats clearly and concisely of the apparatus and requirements necessary to engage in the delightful pastime of photography, and will be found most useful to amateurs."

LADY'S PICTORIAL.

"'Beginner's Guide to Photography.' (Perken, Son and Rayment, 'Optimus.') Revised and enlarged edition, 6d. It is clear and explicit, quite free from unnecessary and confusing technicalities. I can safely recommend this little work to any of our readers who contemplate taking up photography."

ENGLISH MECHANIC.

"'Beginner's Guide to Photography.' Messrs. Perken, Son and Rayment, of Hatton Garden, have issued a second edition of this useful little work, which has already reached a very large sale."

THE

MAGIC LANTERN:

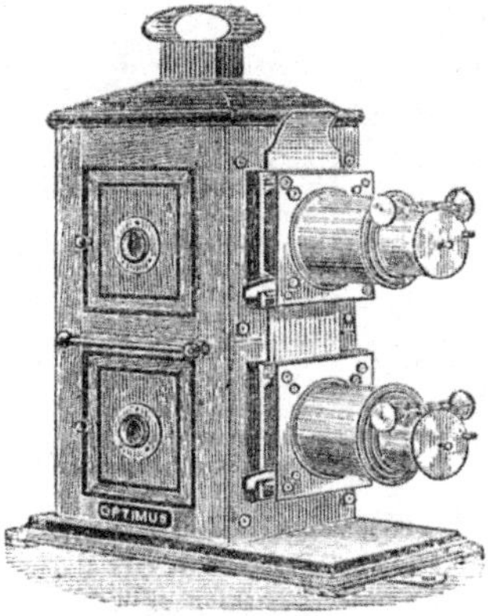

ITS

CONSTRUCTION AND USE.

Copyright.

PUBLISHED BY

PERKEN, SON, & RAYMENT,

99, HATTON GARDEN, E.C.

[FOR PRESS OPINIONS SEE OVER.

PRESS OPINIONS.—"*The Magic Lantern: Its Construction and Use.*"

ILLUSTRATED LONDON NEWS, January 4th, 1890.

"For the entertainment of young folk at Christmas holiday evening parties, the magic-lantern is the most powerful instrument, which has been vastly improved by modern science and skill. An instructive little sixpenny book, written by a Fellow of the Chemical Society, the author of '*The Beginner's Guide to Photography*,' is published by Messrs. Perken, Son, and Rayment, of Hatton Garden. It explains, clearly and precisely, the construction and use of this ingenious optical apparatus, and the advantages of the new and improved magic lanterns, which ought not to be ignorantly or carelessly handled. They seem to be as superior to those which were familiar to the childhood of people now rather elderly, as is the naval artillery of the present day to the ship-guns of Trafalgar."

ILL. SPORTING & DRAMATIC NEWS. January 11th, 1890.

"*The Magic Lantern its Construction and Use.* By a Fellow of the Chemical Society.—Messrs. Perken, Son, and Rayment, of 99, Hatton Garden, publish at sixpence, a little volume, uniform with their treatises on photography and electricity, which deals very lucidly with the subject of the magic lantern. The principles which regulate the construction of magic lanterns, simple and complex, and the methods of illumination, including the preparation of the limelight, are detailed in a way to enable anyone with the most moderate aptitude for scientific matters to master the subject without difficulty."

MORNING POST, January 18th, 1890.

"Messrs. Perken, Son, and Rayment publish a little manual on '*The Magic Lantern, its Construction and Use.*' It explains the numerous improvements which have recently been made in this popular optical instrument, as well as the operator's duties while exhibiting the pictures. Many other matters connected with the use of the lantern, whether for pleasure or educational purposes, are included.

ENGLISH MECHANIC, January 1st, 1890.

"'*The Magic Lantern.*'—Messrs Perken, Son, and Rayment have issued a cheap and useful manual of the magic lantern, explaining the principle of its construction, describing the various forms, and giving directions for its use. All who wish to have a lantern and learn how to manipulate it will find the information required in this handy book.

ARMY & NAVY GAZETTE, January 11th, 1890.

"At this season of the year many amateurs are in search of hints as to how to manage their magic-lanterns. They cannot do better than read an admirable *brochure*, *The Magic Lantern, its Construction and Use*, which Messrs. Perken, Son, and Rayment, of Hatton Garden, have just published. Its instructions are exceedingly simple, lucid, and direct. No amateur following them need hesitate to make use of the oxy-hydrogen or other similar light, and can scarcely fail to succeed with his Lantern.'

AMATEUR PHOTOGRAPHER. January 17th, 1890.

"*The Magic Lantern, its Construction and Use*," is the name of a convenient little hand-book published by Messrs. Perken, Son, and Rayment. It deals with the subject of single lanterns with oil lamps, of the limelight and method of making oxygen, of bi-unial and tri-unial lanterns, and the making and colouring of slides. It also deals with the application of the spectroscope to the lantern, and with the aphengescope. A beginner will find it a valuable help.

BRITISH JOURNAL OF PHOTOGRAPHY, January 10th, 1890.

"RECEIVED.—"*The Magic Lantern: its Construction and Use*," published by Perken, Son and Rayment. A handy manual, a useful feature in which is a priced catalogue of the numerous lanterns and fittings connected therewith sold by the publishers."

THE SPORTING LIFE, December, 28th, 1889.

"'*The Magic Lantern, its Construction and Use*,' is the title of a clever little manual published by Messrs. Perken, Son, and Rayment, 99, Hatton Garden. A perusal of the sound and practical advice contained in its columns should enable one to "build" a lantern of his own, the "tip" being obtainable at the small cost of six pennies."

"FIGARO," January 4th, 1890.

'At this season of the year a little volume called '*The Magic Lantern, its Construction and Use*,' published by Perken, Son, and Rayment, 99, Hatton Garden, will be found extremely useful. It contains a great deal of technical information, and is abundantly illustrated. I do not quite see how anyone who is starting 'the most popular of optical instruments' can do without it. Some excellent hints are given at the close as to the tools which it is necessary for the owner of a lantern to possess.

PRESS OPINIONS.—*"Intensity Coils: How Made and How Used."*

ENGLISH MECHANIC.

"In the preface it warns us that it is not put forth as a scientific exposition of e matter; yet, for all that, many of the explanations are clear and good, and rections for experiments easily to be followed."

ELECTRICAL ENGINEER.

"'Intensity Coils: How Made and How Used.' By Dyer. Sixteenth Edition. London: Perken, Son and Rayment. A book that has reached a sixteenth edition, and which has been before the world for many years, must contain information that is wanted. The Ruhmkorff coil has become one of the most interesting pieces of apparatus in electrical engineering in its comparatively recent development, known as the transformer. This book shows very clearly the historical views held about the coil, and from the amateur constructor's point of view is most valuable, in that it explains clearly how to make and how to use a coil. A great many lecture experiments are described."

CHEMIST AND DRUGGIST.

"'Intensity Coils.' By 'Dyer.' London: Perken Son and Rayment, 99, Hatton Garden. 1s. This is one of those valuable little popular manuals which convey a sound elementary knowledge of an important subject in a concise manner. It describes the principal batteries, deals fully with making coils, and illustrates their use by numerous experiments. There are in addition brief explanations of the electric light, the telegraph, telephone, phonegraph, &c."

ARCHITECT

"'Intensity Coils.' This is the title of one of the publications of Messrs. Perken, Son and Rayment, of Hatton Garden. It is a "beginner's guide to electricity," describing the way to make batteries, bells, coils, electric light, telegraphs, phonographs, telephones, &c. It is already passing through its sixteenth edition, or the 128th thousand. One of the merits of this brochure is that it does not require an expert to understand it."

PHOTOGRAPHIC NEWS.

"Messrs Perken, Son and Rayment, the sixteenth edition of 'Intensity Coils, being a beginner's guide to electricity, describing the way to make batteries, bells coils, electric light, telegraphs, phonographs, and so on.'

PHOTOGRAPHY.

"'Intensity Coils.' Perken, Son and Rayment, 99, Hatton Garden, E.C. This forms one of the firm's well-known publications, and is a sort of beginner's guide to electricity, describing the way to make batteries, coils, lights, bells, telegraphs telephones, phonographs, &c.'

CHEMICAL NEWS.

"The induction coil has now come into such general favour, and is used for such varied purposes as a medical instrument, a means of scientific research, and an amusing toy for winter's evenings, that several treatises have lately appeared, describing the best methods of its manufacture and management. The object of the little work before us is explained in the following manner in the preface:—

"What is an intensity coil? How does it differ from other coils? How is it made? What will it do? These and similar questions are being continually asked, and to furnish intelligible replies to them the following pages have been written It is not a scientific treatise that is here offered to the public, but simply the necessary instructions that those who want to make or use intensity coils desire to obtain."

CITY PRESS.

"'Intensity Coils.' By 'Dyer' (Perken, Son and Rayment). When a handbook, as for instance the one now demanding attention, retains its position for nearly a quarter of a century, few will be inclined to dispute the contention that it is a work possessing considerably more backbone than the generality of such publications. The present book was first published in the year 1867, and it is now in its 16th edition. In its pages will be found, given in an eminently readable form, much that will serve to instruct the young idea concerning the electric light, the telephone, the microphone, and other wonders which, more or less, are associated with the great American inventor, Edison. With all the confidence we displayed when twenty years ago we noticed the first edition, we can heartily commend this handbook to the notice of our readers as a publication 'which will be of service to all persons' engaged in the interesting experiments about which 'Dyer' has so much to say."

WOODFORD FAWCETT & Co.
PRINTERS, DORSET WORKS, SALISBURY SQUARE, E.C.

PRESS OPINIONS.—*"Intensity Coils: How Made and How Used."*

SPORTING AND DRAMATIC.

"'Intensity Coils.' Messrs. Perken, Son and Rayment publish the sixteenth edition of a capital little handbook on this subject by 'Dyer.' Electric lighting, electric bells, the telegraph, the telephone, the microphone, the phonograph, &c., are pithily and comprehensibly explained. The work, which has been revised to date, is published at one shilling so that a practical knowledge of applied electricity up to date is within the reach of everybody."

OVERLAND MAIL.

"Messrs. Perken, Son and Rayment's useful guide to electricity for Beginners, entitled 'Intensity Coils: How made and How Used,' is now passing through its sixteenth edition (128th thousand). It describes the construction and method of using galvanic batteries, while there are chapters on electric lighting, the telegraph, telephone, and Phonograph. The chapters are subdivided into paragraphs, which are headed in large type. To the amateur electrician this work is invaluable."

PICTORIAL WORLD.

"'Intensity Coils: How Made and How Used.' This is the third of a series of scientific primers brought out by the well-known firm of Perken, Son and Rayment, 99, Hatton Garden, E.C. It should prove of great use to all those interested in electricity; it is carefully complied and brought up to date, containing much useful information on electric lighting, bells, motors, the telephone, microphone, and phonograph. Directions are fully given for seventy or eighty instructive and amusing experiments; and we predict a large sale for this, the latest edition."

BUILDERS' REPORTER AND ENGINEERING TIMES.

"'Intensity Coils.' The scientific information is of an advaneed character; but the reader is introduced to it by a preliminary clear and simple explanation of all the rudimentary steps, technical terms, &c. The application of electricity to the various well-known purposes is explained, and the latest projects, and, among these, 'seeing by electricity' are noticed.

MERCANTILE GUARDIAN

"'Intensity Coils.' Sixteeenth edition. By 'Dyer.' London: Perken, Son and Rayment. 1s. Electricity, like photography, is no longer the exclusive pursuit of the professional expert; for the ubiquitous amateur has made it his own, as he has most other things, thus necessitating elaborate handbooks for his particular benefit. The induction coil is much in favour, as it can be applied in so many ways—as a medical instrument, for scientific research, and as part of a drawing-room entertainment. We can advisedly recommend this most useful little handbook, as it contains in a condensed form just the information required. Other branches of the science are treated upon, such as electric lighting, bells and telegraph; the telephone, microphone and phonograph."

INVENTION.

"'Intensity Coils: How Made and How Used,' Perken, Son and Rayment, 99, Hatton Garden. This is a very useful beginner's guide to electricity, although its title hardly gives a fair idea of its contents. It deals with electric lighting, bells, telegraphs, motors, the telephone, and the microphone, while the phonograph is also noticed, on the ground that its invention was an outcome of the investigation of the phenomena connected with the telephone. This being the 16th edition (128th thousand), the fact that the work supplies a public want may be safely assumed."

CIVILIAN.

"'Intensity Coils: How Made and How Used;' also electric light, telephone, phonograph, &c. By 'Dyer.' Messrs. Perken, Son and Rayment, 99, Hatton Garden. This is a little handbook which will be of great use to amateurs engaged in their interesting experiments. It is concisely written, the explanations are clear and good, and it is evidently the work of one who has himself done the work. Students will have to find out what is an intensity coil? How does it differ from other coils? How is it made? What will it do? Besides a mass of other information, including seventy or eighty amusing and instructive experiments. All who have a scientific bent will welcome this little book.

HOME NEWS.

"'Intensity Coils: How Made and How Used.' By 'Dyer,' (Perken, Son and Rayment. 1s.) Describes the manufacture of batteries, bells, coils, electric light, telegraphs, telephones, phonographs, &c., in a crisp, clear, understandable style, which the amateur will value very highly It is not a book of 'pretension,' but no better testimonial as to its merits as a popular handbook to the fascinating subject of which it treats could be mentioned than the fact that the little volume is now passing through its sixteenth edition, or 128th thousand.

PRESS OPINIONS.—"*Intensity Coils: How Made and How Used.*"

ENGLISH MECHANIC.

"In the preface it warns us that it is not put forth as a scientific exposition of the matter; yet, for all that, many of the explanations are clear and good, and directions for experiments easily to be followed."

ELECTRICAL ENGINEER.

"'Intensity Coils: How Made and How Used.' By Dyer. Sixteenth Edition. London: Perken, Son and Rayment. A book that has reached a sixteenth edition, and which has been before the world for many years, must contain information that is wanted. The Ruhmkorff coil has become one of the most interesting pieces of apparatus in electrical engineering in its comparatively recent development, known as the transformer. This book shows very clearly the historical views held about the coil, and from the amateur constructor's point of view is most valuable, in that it explains clearly how to make and how to use a coil. A great many lecture experiments are described."

CHEMIST AND DRUGGIST.

"'Intensity Coils.' By 'Dyer.' London: Perken, Son and Rayment, 99, Hatton Garden. 1s. This is one of those valuable little popular manuals which convey a sound elementary knowledge of an important subject in a concise manner. It describes the principal batteries, deals fully with making coils, and illustrates their use by numerous experiments. There are in addition brief explanations of the electric light, the telegraph, telephone, phonograph, &c."

ARCHITECT

"'Intensity Coils.' This is the title of one of the publications of Messrs. Perken, Son and Rayment, of Hatton Garden. It is a "beginner's guide to electricity," describing the way to make batteries, bells, coils, electric light, telegraphs, phonographs, telephones, &c. It is already passing through its sixteenth edition, or the 128th thousand. One of the merits of this brochure is that it does not require an expert to understand it."

PHOTOGRAPHIC NEWS.

"Messrs. Perken, Son and Rayment, the sixteenth edition of 'Intensity Coils, being a beginner's guide to electricity, describing the way to make batteries, bells coils, electric light, telegraphs, phonographs, and so on."

PHOTOGRAPHY.

"'Intensity Coils.' Perken, Son and Rayment, 99, Hatton Garden, E.C. This forms one of the firm's well-known publications, and is a sort of beginner's guide to electricity, describing the way to make batteries, coils, lights, bells, telegraphs telephones, phonographs, &c."

CHEMICAL NEWS.

"The induction coil has now come into such general favour, and is used for such varied purposes as a medical instrument, a means of scientific research, and an amusing toy for winter's evenings, that several treatises have lately appeared, describing the best methods of its manufacture and management. The object of the little work before us is explained in the following manner in the preface:—

"What is an intensity coil? How does it differ from other coils? How is it made? What will it do? These and similar questions are being continually asked, and to furnish intelligible replies to them the following pages have been written. It is not a scientific treatise that is here offered to the public, but simply the necessary instructions that those who want to make or use intensity coils desire to obtain."

CITY PRESS.

"'Intensity Coils.' By 'Dyer' (Perken, Son and Rayment). When a handbook, as for instance the one now demanding attention, retains its position for nearly a quarter of a century, few will be inclined to dispute the contention that it is a work possessing considerably more backbone than the generality of such publications. The present book was first published in the year 1867, and it is now in its 16th edition. In its pages will be found, given in an eminently readable form, much that will serve to instruct the young idea concerning the electric light, the telephone, the microphone, and other wonders which, more or less, are associated with the great American inventor, Edison. With all the confidence we displayed when twenty years ago we noticed the first edition, we can heartily commend this handbook to the notice of our readers as a publication 'which will be of service to all persons' engaged in the interesting experiments about which 'Dyer' has so much to say."

BRITISH JOURNAL OF PHOTOGRAPHY.

"Messrs. Perken, Son and Rayment have issued a small book—'Intensity Coils; How Made and How Used.' It forms an excellent beginner's guide to electricity, describing the way in which batteries, coils, &c., are made. The previous editions of this work have evidently found much favour, for the present is the sixteenth edition. It contains numerous illustrations."